John Abercrombie, John Adams Library

The British Fruit-Gardener - Art of Pruning

the most approved methods of planting and raising every useful fruit-tree and

fruit-bearing-shrub

John Abercrombie, John Adams Library

The British Fruit-Gardener - Art of Pruning
the most approved methods of planting and raising every useful fruit-tree and fruit-bearing-shrub

ISBN/EAN: 9783337099770

Printed in Europe, USA, Canada, Australia, Japan

Cover: Foto ©berggeist007 / pixelio.de

More available books at **www.hansebooks.com**

THE British Fruit-Gardener;

AND

ART OF PRUNING:

COMPRISING,

The most approved Methods of PLANTING and RAISING every useful FRUIT-TREE and FRUIT-BEARING-SHRUB, whether for Walls, Espaliers, Standards, Half-Standards, or Dwarfs:

The true successful Practice of PRUNING, TRAINING, GRAFTING, BUDDING, &c. so as to render them abundantly fruitful:

AND

Full Directions concerning SOILS, SITUATIONS, and EXPOSURES.

By JOHN ABERCROMBIE;

Of TOTTENHAM-COURT, Gardener:

AUTHOR OF

EVERY MAN HIS OWN GARDENER,

First published under the Name of Tho. MAWE.

LONDON:

Printed for LOCKYER DAVIS, in Holborn.

MDCCLXXIX.

PREFACE.

NUMEROUS are the Treatises written on the present subject; few of them, however, have fairly resulted from *Practice*, and therefore it is little to be wondered, that they have been found erroneous and deficient, and liable to mislead in the most essential particulars.

The following sheets are intended to exhibit the cultivation of fruit-trees on an improved, and, at the same time, on a concise plan, according to the true successful practice of our most eminent Gardeners, in raising and forwarding every species of Wall-Trees, Espaliers, and Standards, to their full perfection of bearing.

With regard to the Method pursued, it may be allowable to observe, that the directions, concerning the respective articles, are not confusedly scattered up and down, in distant parts of the work, nor are the operations belonging to ONE, confounded with those of another: in this Manual the reader will find every Fruit-tree separately and compleately treated of, as to its Botanic Class and Genus, its Growth, Mode of Bearing, Flowers, Fruit, and time of Ripening; its method of Propagation by Grafting, Budding, Layers, Cuttings, Suckers, or Seeds; and the different Stocks suitable to the particular nature of the Tree, with regard to Grafting and Budding. All which are pointed out under the name of that Fruit-tree to which they respectively belong.

As

As Wall and Espalier Trees require singular care in the different operations of Pruning and Training, both in Summer and in Winter, these works are here explained with respect to the different stages of growth, and order of bearing.

Concerning common Standards, the reader will also find ample directions: these do not require, indeed, like Wall-trees, a general pruning; a strict attention, however, as well to their early growth, as to their advanced state, being necessary, they were not to be omitted.

The favourable reception of a former* Work has encouraged it's author to sub-

---

* *Every Man his own Gardener*, seven Editions of which have been printed. — This Work, from a diffidence in the writer, was first published as the production of "Thomas Mawe, Gardener to his Grace the Duke of Leeds, and other Gardeners:" it was

mit to the lovers of Gardening, this his Practice in the Culture of Fruit-trees. Indeed, even after some of the sheets were printed off, his idea, of it's bulk had not extended beyond the limits of a pamphlet: he was not aware, that by printing merely from *Practice*, before he had planned his Treatise in Manuscript, he was liable (as it has now happened) to encrease his Pamphlet to a VOLUME.

He has only to hope, as general utility was his first motive to both these publications, that he shall not be accused of presumption, if he flatters himself that his labours in both will be equally acceptable.

however entirely written by the author of the following sheets: whose claim has since been, in some measure, asserted, by subjoining to the Title-page of the latter Editions, the name of JOHN ABERCROMBIE, to the more popular one of Mr. MAWE.

CONTENTS.

Almond Tree	Page 1
Apple Tree	19
Apricot Tree	56
Berberry Tree	89
Bullace Tree	97
Cherry Tree	107
Chesnut Tree	119
Currant Tree	124
Elder-berry Tree	138
Fig Tree	144
Gooseberry Bush	158
Grape Vine. See Vine.	
Hazel, and Filbert Nut	171
Medlar Tree	179
Mulberry Tree	186
Nectarine Tree	194
Peach Tree	203
Pear Tree	220
Plum Tree	240
Quince Tree	254
Raspberry Shrub	260
Service Berry Tree	272
Sorb Tree	278
Vine, or Grape Vine	284
Walnuts	308

Method

CONTENTS.

	Page
Method of Grafting and Budding	332
Situations, Exposures, and Soils	334
Method of Planting	340
Protection of Blossoms, and Thinning young Wall Fruit	344

Speedily will be published,

THE
Gardeners Pocket Dictionary:
IN A
SYSTEMATIC ARRANGEMENT
OF

All Trees, Shrubs, Herbs, Flowers, and Fruits; with their Uses, Propagation, and Culture, in the British Gardens and Plantations, Green-Houses and Hot-Houses,

Alphabetically digested, and divided into the following Heads;

I. Hardy Trees and Shrubs,
II. Herbaceous Plants,
III. Green-House Plants,
IV. Hot-House Plants.

The Whole comprehending

The General Practice of Gardening, and forming a Daily Remembrancer to Gardeners, Nursery-men, Florists, Seedsmen, and all Promoters of Horticulture; agreeable to the Linnæan System, with the Latin and English Names.

THE ALMOND TREE.

THE Almond is eminent both as a fruit tree, and for ornamenting the shrubbery, &c. early in spring, when in full bloom; and is the original of the ancient genus *Amygdalus*, which, by the botanic characters of the flowers, comprehends also the Peach and Nectarine, as species and varieties of the same family or genus; all of which belong also to the class and order, *Icosandria monogynia*, i. e. flowers containing twenty or more stamina and one style.

The botanists admit but of one real species of the common Almond Tree, which they term

Amygdalus communis, COMMON ALMOND; and is botanically described, *Amygdalus with spear-shaped sawed leaves, having glands at the base: and with flowers mostly in pairs, sitting close to the branches, succeeded by large oval, downy, tough fruit, containing eatable kernels,* comprehending several eminent varieties, distinguished by the following names and properties, viz.

1 Common Almond with a bitter kernel.

2 Sweet-kernelled Almond.

3 Sweet Jordan Almond, large and superior in goodness.

4 Tender

4 Tender-shelled Almond.
5 Hard-shelled Almond.

The fruit, in general, of the Almond tree is valued only for the kernel inclosed in its centre in a stone or nut, it being the only edible part; and is by many greatly esteemed as a choice desert fruit to eat, as well as for various domestic purposes.

Considered as a fruit tree, the Almond, in all its varieties, deserves a place in almost every garden, to encrease the variety of eatable fruits, especially as standard and half standard trees, in which they will also adorn the premises very conspicuously in spring, during their general bloom, and supply us with annual crops of fruit without trouble; ripening in September.

The trees generally assume but a moderate growth, obtaining from fifteen to twenty feet stature, dividing regularly into many branches, and emitting numerous straight shoots annually; the whole forming a large full head, adorned with long spear-shaped leaves, and pale-red flowers, having five petals, producing flowers and fruit, mostly on young wood of a year or two old, immediately from the eyes of the shoot.

They flower early in Spring, before the leaves; arising in a vast profusion all along the young branches at almost every eye, succeeded by large oval downy fruit, consisting of a thick tough pulp, including an oblong nut or stone, containing one kernel, which is the Almond,

mond, and the only esculent part, as before observed; the whole arriving to maturity in September; the outer tough cover splits open, and discharges the stone, with the kernel therein, which is fit both for immediate eating, and to be kept for future use.

The trees are all hardy enough to succeed in any common soil of our gardens, in almost any situation and exposure; and in favourable springs, when their early blossom is not destroyed by frost, they generally produce abundant crops of fruit. However, when designed as fruit trees, they should generally be indulged with a sheltered sunny situation.

They are employed principally as standards and half standards, trained with straight single stems, six or seven feet

feet high for full standards, and four or five for half-standards, branching out at these respective heights, all around into regular heads, planted at twenty or thirty feet distance, and suffered to extend every way nearly according to their natural order of growth; though, for variety sometimes a few trees are cultivated, as dwarfs, for walls and espaliers, and trained in the order of wall trees, &c. nearly as directed for Peaches and Nectarines, and in which they often furnish larger and forwarder fruit than on standards.

Observing, that as this tree bears principally on the young wood, we, in performing the occasional prunings, must carefully preserve a general supply of each year's shoots as succession bearers; as in peaches, &c.

The Method of Propagating them, both as Standards and Dwarfs.

The method of propagating Almond Trees, is both by sowing the stones of the fruit, and by inoculating buds of the approved sorts upon stocks of the Plum, Peach, or Almond; but as the seedlings generally vary to different sorts, the budding or inoculation is the only method by which we can continue the varieties permanent with certainty; and they also sooner attain a fruitful state.

Take, however, both the methods of raising them, as follow:

BY SEED.——Procure a quantity of the stones of the best Almonds that are well ripened in Autumn, and either proceed to sow them in October or November, if quite dry ground, or preserve

serve them in sand, in the dry, till February, preparing for their reception a spot of any good light ground, and drill in the stones two inches deep, in rows a foot asunder; and when the young plants are a year or two old, plant them out in Autumn or Spring, with the spade or dibble, in nursery lines, a yard distance, and half that distance in each row. Train those designed for standards, with tall single stems, from five to seven or eight feet high, by pruning off all side shoots to the height intended; then just top them with your knife at the proper height, either as full or half standards, to force out a set of branches more regularly in the part required to give the head its first regular form; afterwards let them branch out in their own way, and form a full head. But if any are

intended

intended to form dwarfs for walls, &c. they may be topped in their minor growth within a foot of the ground, to obtain branches below to cover the wall and espalier regularly from the bottom, and trained as hereafter directed.

But when any of these seedling Almonds are intended as stocks in which to bud any of the approved varieties, they must not be topped or headed, but trained straight up till after the budding is performed.

By BUDDING or Inoculation.----The propagation of Almonds by budding is effected by introducing buds of the approved varieties into Plum, Peach, or Almond stocks, raised from the stones of the fruit, as directed above in raising the seedling Almonds, and planted out in nursery-

nursery-rows a yard asunder, in order to be trained to the proper sizes for the reception of the buds.

If for standards, may either use low stocks of two or three feet in height, in order for the bud to be inserted near the ground, and the first shoot trained up for a stem; or the stocks may be previously run up to stems the proper height, from five to six or seven feet, to receive the bud near the top, at the height proper to form the first branches of the head—But

When designed as wall or espalier trees, stocks of two or three feet stems are sufficient, as the budding must be performed within half a foot of the ground.

Thus,

Thus, the stocks being ready, the budding is performed in July, in the usual method *(see Budding)* and in the Spring following the buds shoot forth, each one strong shoot attaining two or three feet in length by Autumn, and commences the new tree.

Then the first shoot from the budding is to be managed, as the case requires, to give the trees their first proper form. If such standards that are budded low in the stock, the first shoot must be run straight up to form a stem of proper height; but if the standards are budded at top of the stock, and this forms the stem, the said first shoots should, generally, either the same Summer in June, or in March following, be headed down to six or eight inches; as also those of the dwarf trees, to force out from the lower eyes a supply of three four or more new shoots near the top of the stem, to commence the head in its first regular formation. After this, generally permit the standards to branch in their natural way; and let the dwarfs have their branches trained horizontally to stakes, while in the nursery, afterwards to the wall or espalier.

The trees thus raised, should, when from one or two to five or six years old, have their final transplantation into the garden or orchard, &c.

Planting, and general Culture.

As to planting, the proper season for this is any time from the end of October till March, in open weather; having the trees dug up carefully in the nursery

with

with all the roots poffible, and of which prune off only broken parts, and any ill-placed rambling fhoots of the head; then opening capacious pits for their reception, twenty or thirty feet diftance for the ftandards, and fifteen or eighteen for the wall and efpalier trees, plant them with due care, and a pot of water given to each tree will fettle the earth properly, and promote their rooting. See *Planting*.

Then with refpect to the general culture in the Garden, &c. obferve as follows:

Firft, in regard to the ftandards; they having had their firft fhoots fhortened in the nurfery as directed, and thereby obtained

obtained several well-placed branches near the top of the stem, to adorn the head with a regular shape, should now suffer the whole to branch out freely every way, and only just reform with your knife any very irregular-placed branch, and long rambler, or thin out occasional crowded wood, leaving always the general supply of regular branches entire, and suffered to shoot according to nature.

Next, with regard to the wall and espalier Almonds; they having their first shoots from the budding headed down in the Spring, as already directed, and having obtained three, four, or more regular branches near the bottom, train them horizontally to the wall, equally to the right and left, at full length all Summer; and in Winter or Spring following may prune them to ten or twelve inches, to promote a farther neceſſary ſupply of ſix, eight, or more, regular branches below, training them as above, after which they need not be pruned ſo ſhort, but continue encreaſing the number of horizontals every year, trained five or ſix inches diſtance, till they cover the aſſigned ſpace of walling and eſpalier in a regular expanſion.

And, as the trees thus trained will throw out numerous uſeleſs ſhoots every Summer, they will accordingly require a general pruning and training every Summer and Winter, in common with other wall and eſpalier trees; in order,

both

both to retrench clofe the fuperfluous young wood, all fore-right, and other irregular fhoots of each year, that cannot be trained in. Therefore, going over the trees timeoufly every Summer, rub or prune off clofe all the above evidently ufelefs growths; felecting at the fame time a fufficient general fupply of the regular fide fhoots for fucceffion bearers, and trained at full length till Winter: Then in Winter pruning, chufing a full fupply of the beft fituated of the laft Summer fhoots in every part, let all the others that are not wanted be cut away quite clofe: likewife, at this pruning, cut out part of all the old horizontals or former bearers, down to the beft placed young fhoots they fupport; and the general fupply of young wood, now retained,

retained at four or five inches distance, should generally in this, the Winter pruning, be mostly shortened, more or less; the smaller shoots to eight or ten inches, and the stronger ones to twelve, fifteen or eighteen inches, or two feet long, or more, according to their strength, to promote their producing more certainly a requisite supply of lateral shoots next Summer from the lower eyes, properly situated to train in for succession bearers, as in the Apricot, Peach, and other trees, that bear principally on the young wood, (*See peaches, &c.*)

Then, as soon as ever a tree is thus Winter-pruned, let the whole be directly nailed regularly to the wall, or tied

tied to the espalier, ranging the branches horizontally, and equally to both sides, as before, four, five, or six inches asunder, no where crossing, but all laid parallel or beside one another, at these distances, and fastened along as strait and neat as possibe.

APPLE-

APPLE-TREE.

THE Apple is justly esteemed the best fruit-tree in the world, for the great value of its most excellent fruit, for numerous important domestic uses all the year round, and comprehends varieties without end, all the offspring of one original species, which by the botanists is retained as a species of *Pyrus*, or Pear-tree, their mode of bearing, and botanic characters of the flowers, &c. being exactly similar; and although the fruit differs in external form, yet they perfectly correspond internally; but the Apple, however, was long considered as a distinct family, or genus,

genus, under the title of *Malus*, till, by the laws of modern botany, it was ranged as a species only of the genus *Pyrus* aforesaid; and they both belong to the class and order *Icosandria Pentagynia*, i. e. flowers having twenty or more stamina, and five styles.

There is only one real species of the common Apple tree, comprehending many varieties, and is by the botanists entitled,

Pyrus Malus, i. e. the APPLE TREE, and is botanically described—*Pyrus, with oval sawed leaves, and the flowers produced in close-sitting umbels,* succeeded by large round and oblongish fruit, concave at the base; this being the

the specific distinction of the Apple-tree. It comprises numerous varieties, differing both in the size of the trees in their general growth, as well as in the strength of the shoots, size and colour of the leaves; but more conspicuously in the size, shape, colour, qualities, and times of ripening of the fruit; the varieties of which, no doubt, amount to some hundreds.

However, in respect to the fruit, we shall exhibit a catalogue of only the principal well-approved sorts, and such that are known and readily obtained in all the public nurseries, by the names annexed in the following list, which exhibits also their time of ripening.

JENNETING, or June-eating Apple. *July.*

CODLIN. *July till Sept.*

MARGARET APPLE. *August.*

QUINCE APPLE. *September.*

KENTISH CODLIN, *large. Aug. and Sept.*

SUMMER PEARMAIN. *Aug. and Sept.*

SCARLET PEARMAIN. *Aug. and Sept.*

GOLDEN RENNET. *Sept. and Oct. &c.*

GOLDEN RUSSET. *Sept. and Oct. &c.*

SUMMER REMBOURGE. *Aug. and Sept.*

SUMMER CALVILLE. *Aug. and Sept.*

RED AUTUMN CALVILLE. *Sept. and Oct.*

WHITE AUTUMN CALVILLE. *Sept. & Oct.*

GOLDEN PIPPIN. *Sept. Oct. and Winter.*

AROMATIC PIPPIN. *October, &c.*

WHITE RENNET. *October, &c.*

LOAN'S PEARMAIN. *Sept. and Oct.*

ROYAL PEARMAIN. *Sept. and Oct.*

VIOLET

Violet Apple. *Oct. and Winter.*

Nonpareil. *Oct. Nov. Winter, &c.*

Large Nonpareil. *Oct. and Nov.*

Royal Russet. *October till Spring.*

Rennet Grise. *October and Winter.*

Monstrous Rennet. *Oct. Nov. &c.*

Wheeler's Russet. *Oct. and Winter.*

Kentish Pippin. *October, &c.*

Courpendu, or Hanging Body. *September, October, &c.*

Holland Pippin. *Oct. Nov. &c.*

Kirton Pippin. *Sept. and Oct.*

Winter Greening. *Oct. and Winter.*

Anise Apple. *Sept. October, &c.*

Orange Pippin. *Sept. Oct. &c.*

Winter Pearmain. *October, &c.*

Pile's Russet. *October till Spring.*

Stone Pippin. *Oct. and Nov. till Summer.*

Embroidered Apple. *October, &c.*

Grey

GREY LEADINGTON. *Sept. Oct. &c.*
LEATHER-COAT RUSSET. *October and Winter.*
NONE-SUCH. *September, October, &c.*
WINTER QUEENING. *Oct. Nov. &c.*
POMME D'API. *Sept. Oct. and Winter.*
CAT'S HEAD. *September and October.*
WHITE COSTIN. *September and October.*
LEMON PIPPIN. *October, &c.*

To the above varieties may also be added the *Wilding*, or *Crab-Apple*, and the *Paradise*, or *Dwarf-Apple*,—But

The *Crab Apple*, supposed the original, or parent species of the whole, is valued only principally to raise for stocks, on which to ingraft the cultivated apples; and for the fruit, to make verjuice.

And

And the *Paradise Apple*, which is of very dwarfish growth, is chiefly used only for stocks to graft upon, to form very low dwarf trees, as we shall have occasion to mention.

Thus we may observe, by the foregoing list, the varieties of Apples are very great; and there are still a much greater variety than here mentioned, but those are the best we can recommend; and of which, about fifteen or twenty different sorts, that follow one another successively in perfection, may be sufficient to furnish a supply the year round; however, where there is full scope of ground, a person may indulge his fancy with a more considerable variety, choosing a proper collection both of

Summer and Winter kinds, but most of the latter, for general use; the trees of all which may be had at the public nurseries, in the greatest perfection, both standards, and espaliers, &c. and may be all easily raised by grafting, as hereafter directed.

The Apple tree grows twenty feet high, or more, with a large spreading head, and produces its flowers and fruit upon spurs, short robust shoots, half an inch or an inch long, issuing from the branches of from two or three, to many years old, appearing first towards the upper parts; so the branches must never be shortened in the general course of pruning.

The trees are very hardy, succeed in
any

any common soil of a garden or orchard, or in any situation where it is not very wet, which we should abandon, for they do not prosper in low wet ground; generally allotting them an open exposure, free to the sun and air.

They succeed both as standard trees, half-standards, and in espaliers; but we rarely indulge them with a wall, as they ripen their fruit abundantly well without that assistance. In espaliers, however, we should generally allot a good collection of the best Eating Apples, as in that order of training they commonly attain superior perfection, in regard to size and beauty, than on common standard trees; but all the sorts also acquire perfect maturity on standards;

and it is the common full standard trees, that furnish us with the principal supply of all sorts of Apples for general use, being planted in ranges in the garden or orchard, thirty or forty feet distance, and permitted to branch out every way according to nature; and the espalier trees, planted twenty feet distance, with their branches ranged horizontally to the trellis, always at full length, because shortening those trees would cut away the very first bearing parts, and retard their bearing, especially as they begin bearing first towards the ends of the branches.

The full standards are trained with tall upright stems, six or seven feet high, before permitted to branch out

to

to form the head, and are the proper trees for general culture as standards. And half standards are trained with stems only three, four, or five feet high, then suffered to branch out at that height, and form the head.

But dwarf trees for espaliers, &c. are trained with low stems, only ten or twelve inches high, in order that they may form branches near the ground, to cover the whole espalier from the bottom to top.

And dwarf standards are also trained with low stems, only a foot or two high, for small gardens, and by way of variety.

Method

Method of Propagation.

With regard to the method of propagating Apple-trees, all the different varieties having been first obtained accidentally from seed, and as they rarely come to the same sorts again by that method of propagation, the approved sorts thereof are propagated, and continued permanent, only by grafting shoots of them into any kind of the common Apple, or Crab; stocks raised from the kernels of the fruit, for all common standards and espalier trees; also, occasionally, upon Codlin stocks, raised from suckers, cutings, and layers, to form moderate standards, espaliers, and dwarf standards; and on Paradise stocks to have very

dwarf

dwarf trees, to accommodate small gardens, and for curiosity.

All stocks, raised from any kind of cultivated Apples, are called free stocks, to distinguish them from crab stocks, and from the dwarf stocks of the Codlin and Paradise Apple.

Having observed thus far, let the supply of common stocks for general grafting, for common standards and espalier trees, be raised from the kernels of any sort of Apples or crabs, aforesaid, sowed in Autumn, Winter, or Spring, in beds of light earth, drilled or bedded in an inch deep; *(see Stocks)* they will come up in the Spring; and in Autumn, or Spring following, plant out the

the strongest in nursery rows, a yard asunder; here trained three or four years, or till seven feet high for full, and four or five for half standards, and for espaliers and other dwarf trees, two or three feet high is sufficient; though sometimes standards are also grafted on low stocks near the ground, and the first shoot from the graft trained up the proper height for a stem.

The grafting is performed in March, by whip-grafting for small stocks, and cleft-grafting for larger; cutting the grafts from trees noted for bearing the best fruit, and proceed to ingraft them in the usual method; (*see Grafting*) previously heading the stocks; the standards, at from four to five or six feet height,

height, and espaliers and other dwarfs within six or eight inches of the ground; so inserting the graft at top, one in each stock, they will all shoot forth the same year, each three, four or more shoots effecting the first formation of the head, which you will form for the purposes intended, whether they remain in the nursery, or transplanted into the garden at a year old.

Such, for example, the standards, if furnished with four or more regular shoots in the head, they may either have the said first shoots remain entire, if you would have them form a more erect and lofty head; or, if the present first-shoots are not sufficient to give the head its proper form, may shorten them

them in Spring following to five or six eyes, to force out more branches the ensuing Summer, near the top of the stem, if you design them to form a lower, more full, and wide spreading head; after this, however, permit the whole to branch out every way at full length, and they will thus form a bearing state in two or three years.

And as to the espaliers and dwarf standards, it is generally advisable to shorten all the first shoots from the graft, in March following, to five or six inches, to obtain a good supply of lower branches in Summer, to form the head more full and regular, quite from the bottom; afterwards trained generally all at full length. Those of the espalier trees ranged

ranged equally to both sides horizontally to stakes, displacing any fore-right, irregular shoots that cannot be trained in with regularity: and the regular trained branches being continued always at full length, they in two or three years emit many fruit spurs, and commence bearers. See their *General Culture*.

Thus the young trees, both standards and espaliers, when from one or two to three or four years old, are proper for final transplantation into the garden or orchard, &c. as below.

Planting them in the garden, &c.

As just above observed, Apple-trees of from one or two to three or four years

years old from grafting, having formed good heads, furnished with several branches, are more eligible than older trees for planting; and the proper planting season is any time from November till March, in open weather.

At the proper time let the trees be digged up in the nursery, with all the roots as entire as possible, pruning off only broken parts, and leave the head wholly entire, except just retrenching any ill-placed shoot, not consistent with the general form; then plant them according to the usual method; (see *Planting*) the full standards thirty or forty feet distance, both in the garden and orchard, the half standards not less then twenty five feet asunder, and the

espalier

espalier trees twenty feet distance at least, with their branches ranging the way of the espaliers; but if grafted on free or crab stocks, twenty five feet is not too much; and those upon codlin stocks, not less than sixteen or eighteen feet distance; but those that are on paradise stocks, fifteen feet may be sufficient: the same rules are to be observed in planting dwarf standards.

As soon as planted, let the high standards be properly supported with stakes, and fasten the branches of the espalier trees along to the trellis or stakes of the espalier. *(See Planting.)*

General Culture of Pruning, &c.

First in respect to pruning of standard Apples: they having formed heads furnished

nished with several regular branches, extending each way, by the rules observed in their nursery culture, should generally proceed with the whole at full length, branching out freely every way around to their full extent, and form a full spreading head; and thus they will naturally emit numerous fruit spurs all along the sides of the branches as they advance in length, and bear abundantly without trouble of much pruning; for standards having full scope to branch out fully on all sides, require but only a trifling pruning, probably once in in several years, just to retrench any very ill-placed branch that grows aukwardly across the others, &c. reduce long ramblers and low stragglers, or occasionally to thin such as crowd or confuse the head considerably, and casual

casual worn-out barren branches, and all decayed wood, cutting close, either to their origin, or down to some commodious lower shoot or branch, as it shall seem convenient, and to cut out also all suckers or shoots from the side of the stem, or that rise in the middle of the head. See *Pruning Standards.*

But as to the espalier Apple trees, they being limited to a certain space, both in height and extent sideways, having their branches trained at regular distances, and as they will annually produce many more shoots than are wanted, or can be trained, consistent with the general regularity, they consequently require a general pruning and training every year, to retrench the redundant and irregular

irregular wood, and to train in occasional new supplies where wanted.

In the infant state of espalier trees, our principal care is to procure a good spread of lower branches to furnish a proper basis, as it were, to supply others regulary upward, to cover the espalier equally from bottom to top, with bearers; which depends wholly on the first and second year's pruning, by shortening the first shoots more or less, as directed in the nursery; so that supposing the new planted trees are only one year old from grafting, and their first shoots or head entire, not being shortened in the nursery as directed, it must now be done in the Spring, cutting each shoot to five or six inches,

as

as directed in the nursery culture, to obtain a more abundant and regular supply of branches below, consisting of six, eight, or more, ranging regularly the way of the espalier.

But, if the young trees were previously headed once or twice in the nursery, as there advised, and thereby furnished a proper supply of lower laterals, forming a regular spread of eight or more branches, near the bottom, as aforesaid, we need not practice pruning short hereafter, as having now obtained a good foundation below, they will generally afford a sufficient supply gradually to fill the espalier upward.

In this case it is advisable to train the

the whole now principally at full length, horizontally along to the espalier, for bearers, ranging an equal number to both sides, five or six inches asunder, tying the branches close to the trellis; and, as they advance in growth, still continue them at full length; for shortening would retard their bearing and force out useless wood; so only shortening an occasional shoot, where wood is wanted to fill a vacancy, being either pinched down to a few eyes early the same Summer, or cut short in Spring, to force out three or four shoots below.

In this manner continue increasing the supply of horizontals or bearers annually upward, at the above distance, one above another, till they by degrees cover the espalier regularly to the top, at the same time being careful to retrench all superfluous and irregular wood of each year, as directed below, taking them off always quite close, fastening the regular branches always straight and horizontally to the espalier.

Thus, the branches being trained always at full length, they will soon all emit numerous fruit spurs, and bear plentifully, and will continue increasing the supply of fruit spurs as they advance, and the same branches continue improving in bearing for many years.

Their General Pruning in Espaliers.

Remembering, as the same branches continue fruitful many years, no general

ral supply of wood is wanted annually, after the trees are fully trained, as in Almonds, Peaches, and Nectarines, &c. which bear only on young shoots of a year old, but only now and then a shoot retained here and there, as hereafter explained.

Remark, likewise, in the general pruning of these trees, never to shorten the branches, because, as they bear first towards the extreme parts, and encrease the number of fruit spurs as they advance in length, that, if shortened, would cut away the very first bearing parts, and every shortening would retard their bearing two years, besides promoting a great luxuriancy, and in the places where fruit spurs would have likewise appeared,

appeared, send forth numerous strong useless wood, and no fruit.

Espalier Apple trees in general require pruning and training every Summer and Winter to preserve the regular form, &c. by retrenching the redundant, and training in the proper growths.

The Summer pruning is requisite in order to reform the irregularities of the same year's growth, so should begin to go over the trees in May, or early in June, before the shoots of the year are much advanced, and carefully prune out all the fore-right, and evidently superabundant and unnecessary young shoots, retrenching them quite close; which, if begun early before the

shoots

shoots become woody, may be effected with the thumb, otherwise must use the knife; being careful to leave the terminating shoot of every horizontal or bearer entire, where there is room to extend them, and reserve here and there a well placed strong shoot towards the lower parts, at full length, especially where there is any apparent defect or vacancy, or if none, it is proper to leave some good side shoots in different parts, till Winter pruning, in case of any unforeseen vacancy; and if then not wanted, are easily retrenched: training at the present the whole reserved supply close to the espalier; and continue them so with the greatest regularity all the Summer, by reviewing them frequently during the Summer's growth,

growth, to retrench all after-shoots, reform occasional irregularities, and to train in the reserved shoots as they advance in length.

It is highly advisable to begin the Summer pruning, &c. of espalier trees, early in the season: as the work is then not only more easily performed with expedition and truth, either by rubbing off the buds when quite young, or by pruning with the knife the more advanced shoots, but it also contributes exceedingly to the prosperity of the fruit in general, as well as to the beauty and regularity of the trees during their Summer's growth.

The Winter pruning may be performed

formed any time from the fall of the leaf, in November, until March, and consists of a general reform, where necessary, both of pruning and training; previously remarking, as before observed, that, as the same bearing branches remain fruitful many years, they must be every where preserved while they support a good bearing state; and and only introduce a supply of new bearers occasionally; as below.

So that in the operation of Winter pruning espalier Apples, we must examine both the general bearers, and the occasional reserve of the last Summers wood; that if any worn-out, or naked old wood, not furnished good fruit-spurs, occurs, now is the time to retrench it, by pruning down to some more eligible

lower

lower branch, or young shoot, properly situated to supply its place, or where the general branches are too much crowded, should also at this season thin out the most irregular, and cut out dead wood; at the same time, if there are any vacancies, retain some contiguous side shoots reserved in Summer, to supply the deficiencies, or if any good shoot offers towards the bottom contiguous to very old branches, it may be trained up between, to a bearing state, to be ready when wanted; all other young shoots retained last Summer, not now wanted for a supply of wood, either to increase the requisite spread of bearers, or to fill any casual vacant space, must be cut

cut clean out close to their origin, leaving no stump, being careful, however, to preserve the terminating or leading shoot of every horizontal or bearer generally entire, and extended in length as far as the allotted space will admit;

Likewise preserve with the greatest care all the fruit spurs in every part, except any are become very old and barren, or project too considerably fore-right, which retrench close;

And all clusters of large ragged useless spurs, formed by the remaining stumps of shortened shoots, left by injudicious pruning, should now be pruned close off, leaving no stump or spur

spur but the proper fruit spurs, naturally produced;

For, in retrenching the superfluous and bad wood, we should always cut quite close, leaving no stump, as is too commonly practised, whereby, they shooting out at every remaining eye next Summer, crowd the tree with innumerable useless branches, occasioning great trouble to retrench them, which, by unskilful pruners, are stumped off again to an inch or two long, continuing the same practice from year to year, forming, at last, those large clusters of unsightly useless spurs we often see in ill-managed espalier trees.

After the general reform of Winter pruning, let all the branches be regularly ranged in their proper horizontal poſition, at equal diſtances, cloſe to the trellis of the eſpalier, as ſtraight as poſſible at their full length, if room permits, tying them all in neatly with ſlender oſier twigs, &c. See *Eſpalier Trees.*

Gathering the Apples, &c.

Apples arrive to perfection for uſe, in different varieties, from July or Auguſt, until the end of October or beginning of November: the Summer and earlier Autumn kinds, attaining maturity in Auguſt and September, fit for

for use, immediately off the tree, and do not keep long, especially the earlier kinds, but the Winter Apples, which do not attain full growth till October, aforesaid, being then properly gathered, keep good many months and improve in perfection as they lie in the fruitery.

But all the late Autumn and Winter Apples, particularly, should be permitted to have their full growth on the tree till October, some to the beginning and middle, others till towards the latter end of the month, if the weather continues mild and dry.

To know when the Apples are arrived to maturity on the trees, should try if they quit their hold easily on being turned gently up; or that they naturally drop from the tree in any great abundance; or some sorts by changing colour and emitting a fragrant smell; at which tokens of perfection they should be gathered, both Summer and Winter kinds.

They should generally be gathered in dry weather; and all those intended for long keeping should be gathered carefully, by hand, without bruising, carrying them directly into the fruit room, &c. disposed in heaps, each sort separately, to remain a week or two to sweat and discharge the watery juices; then wiped dry, and put up in the different divisions and shelves of the fruitery,

fruitery, and in boxes, or hampers, &c. and then cover the whole closely with clean dry straw, a foot thick, to exclude the moist air as much as possible, whereby they will keep much longer in perfection.

Generally keep the door and windows closely shut, for the less the external air is admitted, the better the Apples will keep.

The APRICOT TREE.

"THE Apricot is one of the most excellent stone fruits, a species of the *Prunus*, or *Plum-tree*, but formerly ranked as a distinct genus, by the title of *Armeniaca*: however, the characters of its flowers and fruit, agreeing exactly with the *Prunus*, the Botanists have ranged it as a species of that genus, and are both of the class and order *Icosandria Monogynia*, flowers having many stamina and one style."

There is but one species of the Apricot tree, comprising eight or nine excellent

cellent varieties of the fruit, and is named by the Botanists,

Prunus Armeniaca, i. e. ARMENIACA, or the APRICOT TREE, — specifically described *Prunus with nearly heart-shaped leaves and flowers*; having five petals, *sitting almost close to the branches,* succeeded by large, roundish, yellow, pulpy fruit, including a stone or nut,—and comprehends the following varieties, ripening in successive order, from the beginning or middle of July, until the end of August.

EARLY WHITE MASCULINE APRICOT. *Middle of July.*

EARLY RED MASCULINE APRICOT. *Middle and end of July.*

ORANGE APRICOT. *Beginning and middle of August.*

ALGIERS APRICOT. *Early in August.*

ROMAN APRICOT. *Beginning and middle of August.*

TURKEY APRICOT. *Middle of August.*

TEMPLE APRICOT. *Middle of August.*

BREDA APRICOT. *Middle and end of August.*

BRUSSELS APRICOT. *Middle and end of August.*

The first two sorts are small fruit, valued for their early pefection; and the succeeding ones are a much larger, handsome fruit, greatly superior in flavour, and consequently more valuable to cultivate for the main supply; generally planting all the sorts against walls; though

though the two laſt ſorts, which are the lateſt, but moſt excellent Apricots, firm and rich flavoured, will alſo ſucceed in Eſpaliers and detached ſtandards; however, all the ſorts are valuable Summer fruit for different domeſtic uſes, *viz.* When young and green, before the ſtone grows hard, are moſt excellent for tarts, &c. when ripe and gathered, whilſt they remain firm, before they become ſoft and mealy, are the fineſt table fruits of the Seaſon, and when fully ripe, may be converted to an excellent ſweetmeat, being preſerved in ſugar.

The trees generally require training as wall-trees, in this country, againſt a warm wall, in order both to protect their

their early tender blossom more effectually from the attacks of the Spring frosts and cutting blasts, to ensure a more certain and plentiful crops of fruit, and also to improve its growth, and obtain it in the greatest perfection.

Though some of the late sorts succeed tolerably well, trained in espaliers, also in detached half and full standards, as aforesaid.

The Apricot tree grows fifteen or twenty feet high with a spreading head, ornamented with large heart shaped leaves, and numerous reddish flowers of five petals; producing the flowers and fruit principally upon the young wood

wood of a year old, immediately from the eyes of the shoots, and often upon small spurs on the two or three years wood; but mostly upon the young shoots of the former years growth, so that a general annual supply of each years shoots must be every where reserved, as succession bearing wood.

They blossom early in Spring, Febuary and March; and the fruit sets in great abundance in favourable Springs: often affording plentiful supply for thinning off for tarts in May, &c. and the remainder ripens in July and August.

The trees are hardy enough; but as they blossom and set their fruit early in in the Spring, often whilſt ſharp froſts and cutting blaſt prevail, they require the indulgence of a warm Sunny ſituaation, and aſſiſtance of a wall, to defend the tender bloom and infant fruit as much as poſſible from the rigours of the weather; ſo that the trees for the general ſupply ſhould be planted againſt a warm wall or cloſe paling fence, &c. ſome earlier kinds againſt a ſouth aſpect, and others on weſt and eaſt expoſures, to effect a greater variation in the times ripening of the fruit, as well as to obtain it longer in perfection; and thus we may employ all the ſorts, not only as common dwarf wall trees, planted fifteen or twenty feet diſtance; but alſo half ſtandards planted between them, trained alſo as wall trees to

make

make the moſt of every part of the walls, &c. as hereafter directed.

But, as before hinted, ſome may alſo be employed both as eſpalier trees, and as detached ſtandards in open expoſures, to take their chance and encreaſe the variety; and for this purpoſe, the *Breda* and *Bruſſels Apricots*, not bloſſoming ſo early as the others, are generally the moſt ſucceſsful, and often ripen in good perfection, and with peculiar richneſs of flavour.

All the varieties of this tree, ſucceed well in any common good ſoil of a garden, or if of a moderate loamy temperature, either wholly or part, may prove an additional advantage; however

ver any common soil capable of producing good crops of herbage, &c. is eligible.

Method of Propagation, &c.

The propagation of the Apricot tree, being originally from the stones of the fruit, the approved varieties so obtained are encreased, and continued the same by budding them upon any kind of Plum stocks.

Raise the stocks for this purpose, from the stones of any sort of plum, sowed in Autumn too inches deep, and when the seedling plants are a year old, proceed to plant them out, previously shortening their down right tap roots

roots, then plant them in rows a yard asunder, and near half that distance in the lines; and in two years, when about two or three feet high, they will be proper to bud for common dwarf wall or espaliers trees; but for half and full standards, they may either be run up to stems from four or five, to six or seven feet high, and budded at top, or trained only the height as for the dwarf trees, to be budded low, and the first shoot trained to a stem the above height.

Then the budding is to be performed in July or August, procuring cuttings of the young shoots of the year, detatched from the best bearing trees, from which to take the buds:

inserting one bud in each stock within half a foot of the bottom for common wall or espalier trees, and at three or four feet for half, and six for full standards, or as low in the stock as for the dwarfs, and the first shoot from the bud trained up for a stem, as we formerly observed; and having thus performed the budding, the buds remaining dormant till the following Spring, when having headed down the stocks a little above the place of inoculation, each bud will soon after push forth one strong shoot growing a yard or more long by the end of Summer, forming the new trees with good large heads; then in the Autumn, Winter or Spring following, the trees may be transplanted finally into the garden if required; or may

may remain in the nurſery and trained for the purpoſes intended.

Obſerving, in either caſe, that in March following, juſt as the young trees begin to puſh, the whole head or firſt main ſhoot from the budding muſt be ſhortened or headed down clean with the knife to ſix or eight inches, to provide ſeveral lateral ſhoots below the enſuing Summer, to form the head regularly from the bottom, training them horizontally at full lentgh till Winter, (See *their General Culture.*)

Likewiſe, the firſt head of the ſtandards, that were budded at the top of the ſtem, ſhould alſo be headed down in the above manner, to force out lower branches near the place of inoculation, in order to form a more regular spreading head.

But such standards that were budded near the ground, must have the first shoot run up entire to a proper height for a stem, then cut over with the knife, at the height required, to have branches emitted to give the head its first form.

Whilst the young trees remain in the nursery, those designed for walls, both dwarfs and half standards, should have their branches trained accordingly, either as they remain in the nursery lines, by being trained horizontally to stakes, or being previously planted against

against reed hedges, walls, or any close fence, training the branches thereto, being careful to retrench all fore-right shoots, and very rank wood, by rubbing them off early in Summer, and train in all the regular branches at full length till Winter pruning, when they must be shortened more or less, as directed in their *General Culture*.

For as Apricot trees bear principally on the young wood, the shortening that of each year in Winter pruning is necessary, in order to force out a regular supply of shoots more certainly in the proper places, as succession bearers; for the same shoots both produce fruit and succession wood at the same time,

Planting them in the Garden, &c.

When the young Apricot trees are from one to three or four years old, they are of a proper age for planting in the Garden; though if only one year old, with their first heads from the Budding entire, they are rather the most eligible; but they may also be transplanted with good success when several years old, and may be had in the nurseries of some years training, and in a state of bearing, which, by the nursery-men, are called *trained trees*, and which are eligible for persons who are in haste to have their walls covered as soon as possible with bearing trees, as they bear the following season after planting.

The season for planting these trees in general, is any time, in open weather, from the end of October till Spring, as for other hardy trees.

At the proper season, having fixed on their allotted situations; some against a South wall, others on West and East aspects, as we before noticed, proceed to prepare and dig the borders; if for a general plantation, observing, if a poor, or very light, hungry soil, add a quantity of good dung, to be trenched in, or a supply of fresh loam, or a compost of any good earth and dung together, the whole digged in one or two spades deep.

Then having digged up the trees carefully

fully in the Nursery, prune off only any broken or damaged parts of the root, leaving their heads entire for the present, and proceed to plant them along the wall, in the usual method, not less than fifteen or sixteen feet asunder; but if eighteen or twenty feet distance the better, especially if low walls, that in default of height they may have room to extend the branches horizontally; and if tolerably high walls may plant a half-standard in each space between the dwarfs, and trained like them also as wall trees, that whilst the former occupy the lower parts, the standards cover the upper part of the wall: as soon as planted, throw down a pot of water to each tree, if dry ground, especially in Autumn or Spring planting, both to settle

settle the earth about the roots, and to facilitate their rooting afresh, then nail their heads to the wall, and manage them in general as directed below.

Espalier Apricots should be planted at the same distances as directed above, for those against the walls.

And the standard Apricots should be allowed some sheltered, sunny situation in the open borders, or quarters of the kitchen garden, or in the compartments of the pleasure ground, or on grass lawns, &c.

General Culture of Pruning and Training, &c.

The trees being planted where they

are to remain, and if but one year old, with their first head from the budding entire; then, in March following, cut them wholly down to five or six eyes, as directed in the nursery, to promote several lateral branches below, to form the head regularly quite from the bottom, as before mentioned; but if they were previously headed in the nursery, and obtained a proper supply of bottom branches to form the head regularly, they need not now be cut so short, only to eight, ten, or twelve inches, and nailed regularly and horizontally to the wall, about four or five inches asunder.

Then observing in both cases, that as the trees thus shortened will soon after shoot out strongly from all the remaining

ing eyes of each shoot, some regular, and others irregular, we must carefully retain all the regular-placed side shoots, and early in May or June rub off all fore-right and other irregular wood, and very rank luxuriant growths; and, when long enough, train in all the regular shoots close to the wall or espalier at full length all the Summer: and then in the Winter pruning, any time from November till Febuary, it is proper to shorten each of the last Summer's shoots to ten or twelve inches more or less, according to their strength, leaving the lowermost shoots rather the longest, then nail them along horizontally to the wall, equally to the right and left on both sides, five or six inches distance; and thus each horizontal will emit a farther supply of branches the following Summer, still being careful to displace fore-right and other irregular wood, rising in front and back of the branches, as soon in the season as possible, and train the rest at full length all Summer as before advised, unless it shall seem eligible to stop or pinch short any particular shoots in a vacant part in May or early in June, to force out a proper supply of laterals the same year, to fill the vacancy as soon as possible; training them in at full length, as directed for the others, till Winter pruning, when they must be shortened and trained as before; and thus the trees will assume a bearing state when two or three years old.

In

In this manner proceed increasing the number of regular-placed branches, annually arranging equally both ways nearly in a horizontal direction, four or five inches aſſunder one after another, till by degrees they cover the wall or eſpalier regularly, from the very bottom to top, conſtantly retrenching all fore-right, very luxurious, and ſuperabundant ſhoots; both in Summer and Winter pruning, cutting then quite cloſe: being at the ſame time careful, to retain every Summer a plentiful ſupply of the well-placed young wood in every part for ſucceſſion bearers the following year, trained moſtly at full length all Summer as aforeſaid, and in the Winter pruning, if too numerous, thin out the worſt placed of

the

the superabundancy; and shorten the remaining regular shoots, to from about eight, ten, or twelve inches to half a yard or two feet long or more, leaving the strong shoots longest in proportion: and then nail the whole close and regular to the wall at the above mentioned distance. See *their General Pruning*.

Thus, it must be observed, that as we advised the succession young wood to be trained at full length all Summer till Winter pruning, and then shortened: the shortening the shoots in the Winter pruning, of these and most other trees which bear principally on the young wood, is necessary, in order to promote lateral shoots in Summer,

mer, from the lower eyes for next years Bearers, which, if the shoots were laid in at full length, would arise only towards the extreme parts, and leave the tree naked of bearers below.

For the best bearing shoots rise principally on the year old wood, that was trained the Winter before; the same shoots both produce the fruit and a supply of succession wood for next years bearers.

Their General Pruning, &c.

As Apricot trees, trained against wall, and espaliers, annually send forth many superfluous and irregular shoots as well as useful wood, they require

require a general Summer and Winter pruning to retrench the useless growth, and to train in the requisite annual supply of regular young wood for succession bearers.

The summer pruning consists in regulating the shoots of the year only, retrenching the bad and train in the useful, and should be begun in May if possible, or while the superabundant and irregular shoots of the year are so young and tender as to be readily displaced with the thumb, or at least, before the shoots in general advance any considerable length, and cause confusion and disorder; keeping in mind in this pruning, always to reserve an abundant supply of the same years well-placed side

shoots

shoots in every part, trebly more than what may be apparently wanted, not less than two or three on each of the present bearers or horizontals, trained in last Winter, in order to have plenty to chuse from in the Winter pruning for next years bearing: but leave no where more than one shoot from the same eye; thus observing to retrench only all the evidently superfluous young wood, ill-placed and very luxuriant shoots as soon as possible, pruning them all quite close, or in vacant parts, pinch some to a few eyes; and having carefully retained plenty of the best regular side shoots in every part, and always some good shoots advancing from below, train the whole at full length till Winter pruning;

pruning; being careful to review the trees frequntly after this during the growing season, to retrench all after shoots, and reform casual irregularities and to continue the whole close and regular to the wall all Summer.

The Winter pruning of Apricots, consisting of general regulation among both the young and the old branches, may be performed any time from the end of October or November, until February or beginning of March, before the blossom buds are too much advanced; previously unnailing most of the principal branches and shoots, that you may more readily examine the work, and have liberty to use your knife properly, as well as have an opportunity of training the branches,

branches, agreeable to the regulation of the general pruning.

Then proceeding to the pruning, examine the general supply of young wood of last Summer, selecting a sufficiency of the most promising and regular-placed of them in every part, for next summer's bearers; one or two, at least, on each, retaining the horizontals trained in last Winter, and cut out close all the superfluous ones before described, and all small twigs, likewise part of most of the former year's bearers, and any very naked branches, unfurnished with young wood, pruning them down to some eligible lateral branch, or young shoots, to make room to train the requisite supply of young wood with due regularity, cutting off close any lateral twigs arising

ing on the selected young shoots, which, as you go on, must be mostly shortened, more or less, for the reasons before explained, and as below,

For example, the smaller shoots cut to six, eight, or ten inches in length, the middling growths to a foot or fifteen inches, and the strongest shoots to half a yard or two feet long; for the strong or vigorous shoots must not be cut short, which would force out luxurious barren wood; being careful in shortening not to cut below all the fruit-buds, distinguished by their turgid, swelling appearance, from the wood buds, which are long and thin, cutting generally either just above a wood bud aforesaid, or to a double fruit bud, on twin blossom,

in

in order to obtain a leading shoot between, at the extremity next Summer, to draw nourishment to the fruit more effectually.

As you also proceed in the pruning, be careful to preserve all the eligible small fruit-spars aforementioned, rising on the two or three year's wood.

But cut out close all considerable projecting old spurs, all dead wood and old stumps.

As soon as any tree is pruned, let it be directly nailed again to the wall with due regularity, arranging all the branches horizontally as before, four or five inches asunder, straight and close in the neatest manner:

Protecting

Protecting the Blossom, and thinning the Fruit.

As Apricot trees blossom early in the Spring, and are often attacked by cutting frosts, and cold blasts, greatly injuring the embryo fruit, if not occasionally sheltered, it is therefore adviseable to afford the best kinds against walls some protection at that period, either of garden mats nailed up before the trees occasionally, or cuttings of any kind of leafy ever-green stock between the branches, to remain till the fruit is fairly set, or past danger.

Sometimes Apricot trees set more abundant crops of fruit than they can nourish,

nourish, in which case it is proper to thin out the redundancy regularly, when about the size of small cherries or gooseberries, leaving not more than two or three on the smaller shoots, three or four on the larger, and so in proportion.

And those fruit thinned off must not be thrown away, they being most valuable as the first green fruit of the season, for making tarts, &c.

Culture of the Standard Apricots.

As to standard Apricots, they, like most other standard fruit trees, want but little attendance in respect to pruning, or any other culture, for after being headed down the first year, and thereby procured

procured a regular set of several branches near the top of the stem, to form a spreading head; let them afterwards generally branch out, and extend in length every way, according to nature, except just reducing any long rambler, very irregular, or crowded branches, and dead wood, which may be performed any time in Winter,—leaving the general regular branches and shoots wholly entire, and they will naturally afford plenty of young bearing wood, as well as fruit spurs, and in favourable seasons will produce good crops of very fine Apricots.

The BERBERRY TREE,

OR,

PEPPERIDGE BUSH.

THE Berberry is of the shrub kind, and held in esteem both as a fruit shrub for its berries, and as an ornamental shrub for adorning the shrubbery, is an inhabitant of many of our woods and hedges, but has been long admitted a resident of gardens, it producing numerous bunches of beautiful red berries, in much estimation as a domestic fruit to pickle, &c. and effects a beautiful variety as they grow on the trees, which belong to the botanic class and order *Hexandria Monogynia*, flowers having six stamina and one style.

There is but one species cultivated as a fruit shrub, and the Botanists call it,

Berberis Vulgaris, COMMON BERBERRY-TREE described, *Berberis having each flower-stalk sustaining a racemous bunch of flowers*; succeeded by clusters of small, bright-red oblong berries, containing two stony seeds, and comprises the following varieties ripe in Autumn, *viz.*

1. Common Red Berberry with stony seeds.
2. Red Berberry without stone.
3. White Berberry.
4. Black sweet Berberry.

But

But the firſt two varieties, being the red fruited kinds, are the principal ſorts for our purpoſe, and the ſtoneleſs ſort is in moſt eſteem for general uſe, particularly for pickling: the berries of theſe two varieties being of a beautiful red colour when fully ripe, and of an agreeable acid reliſh, are in much eſtimation as a choice and very wholſome pickle: and the bunches of fruit are in great demand as an ornamental garniſh to diſhes when ſerved up to table, ſo that a few trees of each of the red ſorts ſhould be admitted into every good garden, as ſtandards, to produce fruit for the above purpoſes.

The Berberry ſhrub is but of moderate growth, riſing only about ten,

ten, or twelve feet high, armed with thorns, and garnished with small oval leaves and loose bunches of yellow flowers of six petals: producing the flowers and fruit on the sides of the young branches.

All the varieties are very hardy, and will prosper any where in the garden, or orchard, and shrubberry.

When designed as fruit shrubs, they should be generally trained as half or full standards, each with a single stem, four, five, or six feet high, then encouraged to branch out at that height, and form a regular head: and being arranged singly in the garden or orchard &c. fifteen or twenty feet asunder, they

they will produce a plentiful crop of berries fit for use in September and October.

Method of Propagation, and Training.

They are propagated by suckers, layers, and by seed, but the suckers and layers are the most certain methods to continue the varieties distinct, and more certainly the layers.

BY SUCKERS. The shrubs send up suckers abundantly from the root, which dig up in Autumn, &c. with as much roots as possible, and plant them in nursery rows a yard asunder, and trained with single stems, pruning off all side shoots till arrived to four, five, or or six feet height, then may either top them with the knife to force out shoots near together at the top of the stem, to form a spreading head, or permit them to run up and aspire more in height.

By Layers.—Chuse the young branches of last Summer, and in Autumn or Winter, &c. lay them down in the earth three inches deep with the tops out in an erect possition; and by next Autumn they will be rooted, then cut them from the parent plant, and set them out in rows, and managed as the suckers.

By Seed. Sow the ripe berries in Autumn in drills an inch or more deep, they

they will probably moſt of them remain till the ſecond Spring before they come up, giving water in Summer, and when the ſeedlings are a year old, plant out the ſtrongeſt in nurſery rows and train them for ſtandards, as above adviſed.

Tranſplanting into the Garden, and Culture.

When the Berberry-ſhrubs are four or five feet high or more, they may be planted out finally any time from November till March; ſome in the garden, others in the orchard, four or five yards aſunder; ſome alſo in the ſhrubbery, &c.

As to culture, very little is wanted, let them branch out freely at top nearly in their own way, only cutting out cafual rank fhoots, very irregular and crowding branches, or reduce long ramblers, all fhoots from the ftem, and fuckers from the root, fuffering all the regular branches to remain entire, and they will naturally form themfelves into plentiful bearers.

The BULLACE-TREE,

OR

WILD PLUM.

THE Bullace-tree is a species of the *Prunus* or Plum-tree, grows wild in our woods and hedges, but is often admitted into curious orchards and gardens as a fruit tree, to increase the the variety of late stone fruits for the desert, &c. it producing a small stone fruit of the Plum kind, which, when fully ripe, eats with an agreeable acid flavour, and is accounted very wholesome; and the tree being of the *Prunus* tribe, it belongs to the *Icosandria*

dria monogynia; flowers having twenty or more stamina and one style.

There is but one species of the Bullace-tree, furnishing some varieties, and its specific name is

Prunus insititia. The BULLACE-TREE. And is specifically described. *Prunus with spinous branches, oval leaves, hairy underneath, and with the flower-stalks, &c. mostly in pairs*; succeeded by small round Plum-like fruit, consisting of a soft sour pulp; including a stone in the centre; and comprehend the following varieties; ripening in October, *viz*:

1. Common

1. Common Black Bullace,
2. White Bullace,
3. Red Bullace,

This fruit being an inferior sort of late Plum, of a sharp acid flavour, merits culture for its late ripening, after all the other sorts, and if fully ripe, eats with an agreeable tart relish, affording a variety among other fruits, both as they grow on the trees, and in the desert at table; and for making tarts, pies, &c. and for preserving, &c.

The Bullace Tree grows twelve or fifteen feet high or more, having thorny branches, oval leaves, and flowers with five petals; producing the flowers and fruit, both from the eyes of the young wood,

wood, and on small spurs from the sides and ends of the older branches.

They flower profusely in April and May, and the fruit ripens the end of September and in October.

The trees are exceedingly hardy, and a few of each sort, propagated and trained as standards, are worthy of culture; for the variety of thier fruit, distributed any where in the garden or orchard, either as full or half standards, dwarf standards, &c.

Method of Propagating and Training.

This tree may be raised abundantly from the stones of the fruit; but the permanency of the different varieties, is continued only by grafting or budding them upon any kind of Plum or Bullace Stocks; and by which they will also bear sooner, and the fruit will be larger.

By SEED or stones of the fruit,—Having a quantity of the ripe Bullaces in Autumn, sow the stones of them in beds of common earth, two inches deep, and when the seedling plants are one or two years old, plant them out in lines two or three feet asunder, and train each with a single stem, three or four feet for half, and six for full standards, then permitted to branch out into full heads.

By GRAFTING and BUDDING, By either of these two methods, any of the varieties

varieties may be continued distinct, by inserting grafts, or buds of them into Plum or Bullace Stocks, raised from the stones of the fruit, as just above directed for the seedling Bullaces, and planted out in nursery lines a yard asunder, to be trained up to the proper height either for half, or full standards, or dwarfs; are thus to be grafted or budded according to the general method, with grafts or buds of the different varieties, and afterwards trained as other standard trees, such as Apples, Plums, &c. with full branchy heads, (*See Apple Tree, &c.*)

Planting and General Culture.

When the trees, raised by any of the above methods, have formed heads, consisting

consisting of several branches, they are proper for final transplantation, where they are to remain, planted any time from November to March, eighteen or twenty feet asunder.

And as to culture as standards, they require very little, suffering them generally to branch in their own way of growth, except occasionally to reform with your knife, &c. any cross-placed and very crowded branches, and all dead wood. Permitting all the proper branches to extend in their own natural manner, and they will afford abundant annual crops of Bullaces.

The CHERRY TREE.

THE Cherry Tree (*Cerasus*) is famous for producing the earliest ripe fruit, of any other kind of fruit tree, attaining perfection at a season when they prove exceedingly acceptable and refreshing, both for the desert and many culinary purposes; is by the modern Botanists, considered as a species of the *Prunus* or Plum-tree; though was long distinguished as a separate *genus* by the title of *Cerasus*; and belongs to the class *Icosandria* and order *Monogynia*, flowers having twenty or more stamina, and but one style.

There are two species of the Cherry cultivated as fruit-trees, viz. *Common Cherry-*

Cherry-tree, and the wild black and red Cherry; diſtinguiſhed by the Botaniſts as follow:

Prunus Ceraſus, i. e. CERASUS, or COMMON CHERRY-TREE, — deſcribed *Prunus with oval-ſpear-ſhaped ſmooth leaves: and with the flowers growing in umbels fitting almoſt cloſe*, — ſucceeded by cluſters of large Cherries, having a ſoft juicy acid pulp, with a ſtone or nut in the centre: and conſiſts of the following varieties, ripening from May till Auguſt, *viz.*

KENTISH, or COMMON CHERRY. *End of June and July.*
EARLY MAY CHERRY (Small.) *May and beginning of June,*

COMMON MAY DUKE CHERRY. *End of May, and in June.*

ARCH DUKE CHERRY, *End of June and July.*

WHITE HEART CHERRY. *June and July.*

RED HEART CHERRY. *June and July.*

BLACK HEART CHERRY. *End of June, and in July.*

AMBER HEART CHERRY. *July and August.*

BLEEDING HEART CHERRY. *Middle or end of July.*

OX-HEART CHERRY. *Middle and end of July.*

LUKEWARD CHERRY. *End of July.*

HERTFORDSHIRE HEART CHERRY. *July and August.*

HARRISON

HARRISON DUKE CHERRY. *July.*
CARNATION CHERRY. *End of July.*
CROWN HEART CHERRY. *July.*
MORELLO CHERRY. *August and September.*

The above catalogue comprises the principal varieties of the common Cherry, known and cultivated in the public nurseries, by the names here annexed; and of which, the best bearers are, the Kentish, all the Dukes, the Lukeward, Hertfordshire and Morello; but most of the Heart Cherries being strong growers, generally bear more sparingly than the other varieties.

Second Species.

2. *Prunus Avium. The Bird's Cherry,* or WILD CHERRY TREE, — *Having*

P 2 *oval*

oval spear-shaped leaves, downy underneath, and with the flowers in close-sitting umbels; succeeded by small round Cherries of a bitterish flavour, comprehending the following varieties, ripening in the end of July, and in August.

> COMMON SMALL BLACK WILD CHERRY.
> COROUN, OR LARGE BLACK WILD CHERRY,
> SMALL RED WILD CHERRY,
> LARGER RED WILD CHERRY,

This second species of *Wild Cherry*, grows wild in woods, and hedges in England, &c. and is often admitted into gardens and orchards, for the variety of its fruit, which effects an agreeable succession;

cession; and by many, much admired for its peculiar bitterish relish: But the Coroun being the largest, and finest fruit, is superior for general culture.

The two species of Cherry-trees differ in growth and magnitude: The common or garden Cherry, grows only about fifteen or twenty feet high, and the second forty or fifty, with a more erect and lofty head; adorned each with spear-shaped leaves, and numerous clusters of white, four-leaved flowers, in April and May, succeeded by the Cherries ripening, in the different varieties, from May till August, or September.

Their mode of bearing is both on the

the young year-old wood, immediately from the eyes of the shoots, and on the older branches, principally upon short spurs issuing, first towards the extreme parts, then gradually along the sides; the same wood continuing fruitful several years, only wanting a renewal of young occasionally, as any branch becomes very old and barren; nor must the shoots or branches be shortened in the course of pruning, as it would destroy the first bearing parts, and promote much lateral wood, and but few fruit spurs.

Though the early May, and the Morello Cherries particularly, generally bear the most abundantly on the young wood, and should always retain

a more

a more plentiful supply of each year's shoots, as succession bearers.

All the sorts of Cherry-trees succeed equally well, trained both as wall-trees, espaliers, and as full and half standards, and occasionally as dwarf standards, for variety, and for forcing; and are all very hardy, prosper in any common fertile soil, and open exposure, in a garden, or orchard, &c. Plant the wall and espalier-trees fifteen or twenty, and the standard thirty feet distance.

As the common Cherry-tree produces the largest and finest fruit, we should cultivate principally the several varieties thereof for the general supply, both for wall, espalier, and standard trees; chusing as wall and espalier-trees,

trees, some of the early May, but more plentifully of the Dukes and other large kinds; some for South walls, for the early supply in May and June, others on West and East walls for succeeding crops, and some also on North walls, to continue the succession till August and September; but the Morello is the most commonly assigned to the northern aspects, though it highly deserves a southerly exposition to improve its flavour; and for standards, may plant any of the sorts, though should generally allot a good share of the Kentish, Dukes, Lukeward, and Hertfordshire Cherry, Black and White Hearts, and some Morello Cherries.

But

But the Wild Cherry in its different varieties, muſt not be omitted in the collection, trained principally as ſtandards, in the garden and orchard; but the ſmall black and red kinds, are alſo often planted, to adorn avenues and parks, and arranged in hedgerows around the boundaries of fields, &c.

Method of Propagation.

Cherry-trees are propagated by grafting or budding ſhoots and buds of the deſireable varieties upon ſtocks, either of the Wild Cherry, as being the hardieſt and ſtrongeſt grower, or on any kind of Cherry ſtock the moſt eaſily obtained; raiſed from the ſtones of the Cherries, as directed for the Apricot, &c.

Performing the grafting in Spring, and the budding in Summer, near the ground, for wall and other dwarf trees, and at several feet height for standards; the grafts will shoot the same year, and the buds in spring following, each forming their first heads by the ensuing autumn, when the young trees may either be transplanted into the garden, or remain longer in the nursery, training them, in either case, for the purposes intended, as directed for the Almonds, Apples, and Apricots.

Final Planting, and Culture, &c.

All sorts of Cherry-trees may be planted where they are to remain, either

either when only one year old, with their firſt heads entire, or from two or three, to five or ſix years growth, any time from November till March; previouſly when digged up, prune away broken parts of the root, and any ill-placed ſhoot of the head, leaving all the reſt entire; then plant the wall and eſpalier-trees, fifteen or twenty feet diſtance, and the ſtandards thirty, or more, if for a full plantation.—

Then in reſpect to general garden culture, take the following hints—And

Firſt of the wall and eſpalier Cherries,—that if new-planted, one year-old trees, having their firſt ſhoots entire, head them down in March to a few eyes, to promote lower branches;

ches; but if headed in the nurfery, and furnifhed with feveral branches below, train them to the wall, &c. moftly at full length, arranging horizontally to both fides, four or five inches afunder; and in Summer, may pinch fhort young fhoots of the year, to procure a further fupply of horizontals; and thus continue encreafing the branches annually, to cover the wall regularly upwards; being careful to rub or prune off all fore-rights, and other irregular growths, and fuperabundant fhoots, training the regular fupply ftrait, and clofe to the wall and efpalier at the above diftances, always at full length, as far as they have room: And they will thus naturally

emit

emit numerous fruit spurs, and bear abundantly in two or three years.—

And, as to general pruning, &c. continue the same branches as long as they remain fruitful, pruned, as below, Summer and Winter.—Always commence the Summer pruning in May or June, to displace all useless growths of the year, such as all fore-right growers, and all apparently superfluous or unnecessary shoots, retaining a supply of well-placed lateral ones till Winter pruning, training them in at full length.—And in the Winter pruning, examining the main branches or general bearers, if any worn out, naked, or dead ones appear, cut them out, and retain young wood in their place; at the same time select occasional well-placed

placed last Summer's shoots in vacancies, advancing for bearers, and a terminating one to each main horizontal, and cut out close all the superabundancies, being careful to preserve all the fruit spurs in every part; then as soon as pruned, nail in all the branches regularly as before, four or five inches asunder at full length, as formerly advised.—

Culture of the Standard-Cherries.

As to the culture of the Standard Cherry, that being trained to the proper form in the nursery, as directed for Apples, &c. and thence planted out fully where their heads have full scope to grow, they afterwards require very little pruning, only occasionally in Winter,

Winter, to regulate any very crowded, and irregular branches, and cut out dead and cankered wood; but otherwise permit the general branches to proceed in their natural growth, as they will soon furnish numerous fruit-buds their whole length.

The CHESNUT-TREE.

THE Chesnut (*Castanea*) is a lofty tree, a species of the *Fagus* or Beech, but formerly a distinct genus by the title of *Castanea*, is of the nut-bearing kind, and cultivated occasionally as a Standard fruit-tree, for variety in orchards, avenues, Parks, Lawns, and out-grounds, &c. belongs to the class and order *Monoecia Polyandria*, i. e. male and female flowers on the same plant, and the males having many stamina, and the females three styles.—

There is but one species of the Chesnut-Tree, called by the Botanists,

Fagus

Fagus Castanea, i. e. CASTANEA, or the CHESNUT-TREE described, *Fagus, with spear-shaped sawed leaves, naked underneath*; and with flowers produced in long Catkins, succeeded by large prickly capsules, containing two or more nuts, consisting of the following varieties, ripening in September; *viz.*

MANURED, or LARGE SPANISH CHESNUT,

WILD or SMALLER CHESNUT.

The fruit of the Chesnut-tree, though not so valuable as many other sorts, forms a variety in Autumn, and Winter; and is by many much esteemed for roasting, in which it eats very tender and palatable.

The Chesnut-tree grows forty or fifty feet high, branching widely round, forming a regular head, ornamented with large elegant lanceolate leaves, and small flowers, without petals, collected in amentums, or strings, at the sides of the younger branches; the females becoming large round prickly Capsules, inclosing the chesnuts, arriving to maturity in Autumn.

It is a hardy tree, grows freely in any common soil, and open exposure; and merits admittance in our fruit-tree collection, especially in extensive grounds, trained as full standards, to plant on the boundaries of orchards, or in parks and avenues, arranged in concert with Walnuts, &c. thirty or forty feet, or more

more distant—in which, when advanced to a tolerable large growth, they will bear abundance of Chesnuts, sometimes little inferior to those we receive annually from Spain and Portugal, &c.

Method of Propagation.

This tree may be raised abundantly from the nuts, and occasionally by grafting, to continue the manured sort distinct.

By the Nuts,—Procure a quantity of well-ripened, plump, sound chesnuts, English or foreign growth, in Autumn or Winter, from the Seedsmen or Orange merchants, and preserve them in sand till February, then planted in drills two or three inches deep, they will come up in six or eight weeks;

and when one or two years old, plant them out in nursery rows, a yard asunder, and here train them with strait clean stems, six or seven feet high for full standards, pruning of all laterals below, and leave the leading shoot entire, permiting them to branch out at the above height, and form full heads, only just retrench any very irregular or rambling growth at first, to preserve a little regularity.

By Grafting.—By this method of propagation we can more certainly continue the manured, or large Chesnut permanent, ingrafting shoots thereof into Chesnut-stocks, raised from the nuts as above, and trained up to high standards, as directed for the seedling Chesnuts.

When

When the trees raised by either of the above methods, are six, seven, or eight feet high, may plant them out finally, where they are to remain, in the places and distances before-mentioned.

And as to future culture, let them branch out mostly in their own way, except retrenching occasionally any very cross-placed and rambling growths.

CURRANT-

The CURRANT TREE.

THE Currant Tree *(Ribes)* is of the shrub and berry-bearing kind, and the most valuable of our fruit-bearing shrubs, for the usefulness of its fruit, which proves cooling and refreshing to eat in the heat of Summer, and excellent for various culinary purposes; and by the rules of botany, comprehends the Gooseberry as a species of the same family *(See Gooseberry.)* and belongs to the Class and order *Pentandria Monogynia*, flowers having five stamina and one style.

There are two species, and are described by the Botanists, as below.

Ribes

Ribes Rubrum, RED CURRANT-TREE. *Having branches without thorns, and plain flowers in smooth pendulous clusters:* Succeeded by hanging bunches of red and white berries, in the different varieties; ripening in June and July, *viz*:

COMMON RED CURRANT
LARGE DUTCH CURRANT
LONG-BUNCHED RED CURRANT
CHAMPAGNE LARGE PALE-RED CURRANT
WHITE CURRANT
LARGE WHITE DUTCH CURRANT

Ribes Nigra, BLACK CURRANT-TREE.—Having *thornless branches, and oblong flowers in hairy clusters*; Succeeded by loose bunches of larger black berries of of a rank flavour, comprising but one useful variety, ripening in July, *viz.*

COMMON BLACK CURRANT.

These shrubs grow six or seven feet high, dividing low into many branches, forming bushy heads, adorned with tri-lobated leave and strings of small greenish flowers, of five petals; succeeded by the bunches of berries; attaining perfection from June and July, till September; and their order of bearing is both on the young and old wood; often immediately from the eyes of the young shoots, but more plentifully upon a sort of spurs or snags arising on the sides of the older branches: and the same wood continues fruitful several years.

The

The Red and White Currants are the moſt eligible for general culture, both as deſert fruit to eat, and for many uſeful domeſtic purpoſes; ſo that great plenty of the buſhes ſhould be admitted into every garden.

But the Black Currant is more in eſtimation for medical uſes than for eating; but is very wholeſome, and ſhould be admitted in the collection, in moderate quantity.

All the ſorts of Currant buſhes are trained both as common buſhy ſtandards, with ſtems a foot, or half a yard high, branching out above into buſhy heads, to arrange in the open quarters of the kitchen garden for the general
planta-

plantations; and occasionally as flat or fanned standards to range espalier-ways in narrow borders, &c. likewise in fanned dwarfs against walls and espaliers, to obtain the fruit in greater perfection and earlier and later in the season, by having them in different expositions.

Method of Propagating and first Training.

All the sorts of Currant bushes are most easily and expeditiously propagated in abundance, both by Suckers, cuttings, and Layers, and raised to a bearing state in two years.

By Suckers—They send up Suckers abundantly from the root every Summer, commencing proper plants by Autumn; when, or in Winter or Spring, dig them up with as much root as poffible, and prune off long, weak, or crooked tops, to twelve, fifteen or eighteen inches length; then plant them in nurfery rows, or the tall ftrong Suckers at once where they are to remain; training the Standard bufhes with fingle ftems, by trimming off all lateral fhoots a foot or half a yard high, then permitted to branch out at top, and form regular heads, keeping the branches five or fix inches afunder, not fhortening the fhoots, (excepting any long rambler,) til lthe head is arrived to the intended height; obferving, the heads may either be permitted to grow convexly or full in the middle, or concave or hollow, by pruning out the central branches, so as to dispose the outward ones circularly around at regular distances.

Others may be encouraged to branch out near the ground in order to be trained in a fanned manner, both for fanned standards, and for walls and espaliers, by cutting away all projecting shoots, and retaining only such as arrange the way of the row, espalierways.

By Cuttings—chuse cuttings of the strong young shoots in Autumn, Winter or Spring, from ten or twelve inches to half a yard long or more, and plant them with a dibble, in any shady border,

in

in rows a foot asunder, each one third into the ground, training them as directs for the Suckers.

By Layers,—In Autumn or Winter &c. lay down any of the lower branches, three inches deep in the earth, with the tops out, they will root freely, and in Autumn following plant them out in rows, and managed as the Suckers.

Final Planting &c.

Currant-trees from two, to three or four feet high, having tolerable branchy heads, are of a proper size to plant out for good; performing it any time from November till March.

Planting the different standard bushes, some in single ranges around the large quarters of the kitchen garden, eight feet in the row, others cross ways; the same distance, to divide large plats of ground into breaks or compartments, thirty or forty feet wide or more; and may likewise arrange some in continued plantations, six feet in the rows, and eight or ten between the ranges:

And for wall and espaliers, plant a few against south walls, &c, for early Currants, and a larger supply on west and east exposures, and plenty on north walls, for general later crops; and some in espaliers, &c. arranging them ten feet distance, and their branches trained either horizontally or ascending, as convenient.

venient, or as room permits, five or six inches asunder, mostly entire, till extended to their limited bounds.

General Culture.

With regard to general culture, let the common Standard Currants be generally continued with single stems, by clearing away all lateral shoots below, and suckers from the roots; and the head kept regular, either convex or full in the middle, or concave or hollow, as formerly explained, with the branches kept five or six inches distance, retaining the same branches several years as bearers.

In Summer, if numerous shoots arise, may go over them with your knife, and trim

trim out close the most irregular and crowding, to admit the Sun to the fruit; reserving a moderate supply of regular ones at full length till Winter.

And in Winter, the bushes will require a more general regulation: that, if the last Summer's shoots remain too abundant, prune out close all the superfluous and irregular, reserving only some occasional regular shoots, advancing in casual vacant parts below, or to supply the places of bad or dead wood, and a terminating shoot to every branch, or when advanced too long, prune it down to such a shoot, or to a lower branch, having one for its leader; likewise now retrench any casual irregular branch, worn out bearer, and decayed wood,

wood, long rambler, or very crowding growths, pruned either to their origin, or down to any more eligible branch or young fhoot; carefully preferving all the natural fruit-fpurs and bearing fnags, and cutting out decayed ones: and then, may either fhorten the upper leading fhoots, more or lefs, to continue the head of a moderate ftature, if required, or permit them to remain moftly entire, and afpire to their natural height and extent, if not limited to room.

Culture againſt Walls and Eſpaliers, &c.

As to the fanned Currants againſt walls and eſpaliers, &c. continue the branches trained four or five inches afunder,

afunder, either horizontally, or afcending, as room admits, extended moftly entire, till advanced to the extent of their utmoft limits.

In Summer prune out all projecting fore-right, very irregular, and fuperfluous fhoots; and train in only fome regular fide and main leading fhoots entire, till Winter.

Then, in Winter pruning, felect occafionally fome well-placed young fhoots in vacancies below, advancing to a bearing ftate, and, according asworn-out or decayed branches occur, they fhould now be pruned down to fome lower young wood: all young fhoots, not now wanted, muft be cut out clofe,
pre-

preserving all the bearing spurs, and cutting out decayed stumps; then either training in the branches entire, where there is room enough to extend them, or in default thereof, shortening the shoots, more or less, as the Case requires;

Of the Fruit.

To preserve Currants long in perfection: in July or August cover some of the bushes both in standards, and against walls, &c. with mats and nets, to shade them from the Sun, and defend them from the birds, whereby a succession of good Currants may be continued from June or July, until October, both for the fruit desert, and culinary preparations.

The ELDER-BERRY-TREE.

THE Elder-tree *(Sambucus)* is of the berry-bearing kind, and merits a place in the fruit-tree collection, for the sake of the berries, for making that excellent cordial liquor, called Elder Wine, which is its chief value as a fruit tree, the raw berries being very unpalatable to eat; and belongs to the Class and Order *Pentandria Trigynia*, flowers with five stamina and three styles or stigmas.

There is but one species of Elder proper for the fruit-tree collection, *viz.*

Sambucus

Sambucus Nigra, BLACK *or* COMMON ELDER-TREE — Defcribed, *Sambucus*, *with a tree-like stem, and cymose, five-parted umbels of flowers*, fucceeded by large umbellate clufters of black, and other coloured berries, in the different varieties, ripening in Autumn, *viz.*

COMMON BLACK-BERRIED-ELDER,
WHITE-BERRIED-ELDER.

The Elder-tree grows twenty or thirty feet high, with a fpreading head, garnifhed with winged leaves of two or three pair of lobes, terminated by an odd one, and large broad umbellate clufters of fmall-white, five-parted flowers, in Summer, at the end of the branches,

branches, succeeded by large bunches of small black berries, ripe in September.

It is the common black-berried Elder we principally recommend as a fruit-tree, for its berries to make wine, and for which the trees demand a place in our collection, trained generally as common standards, or some occasionally in rough hedges; especially as they will grow freely in any soil and situation, in out grounds; such as the verges of orchards, &c. hedge-rows, sides of banks, or ditches of water, or any waste premises, obscure corners, or moist situations; either in detached standards, or in continued hedges; and will produce abundant crops of berries annually, well worth our notice for the purpose aforesaid.

Method

Method of Propagation, &c.

This tree is easily raised from cuttings of the young shoots, and occasionally by seed.

By Cuttings.—In Autumn, Winter, or Spring, cut off a quantity of the best ripened robust shoots of last Summer, in lengths from half a yard to five or six feet, and planted, either at once where they are to remain, or in nursery rows, a yard asunder, introducing each cutting near two thirds, or almost half way, (the longer ones) into the ground, with a long dibble, or with a stake or iron crow for the longest sets; and, as they advance in growth, train those designed for standards

standards with clean single stems, five or six feet high, then let them branch-out with full heads; and permit the hedge plants to grow rough nearly from the bottom, only trimming up the sides little, and rambling shoots.

By Seed.—In Autumn procure a quantity of the ripe berries, and sow them any where in drills an inch deep, and when the plants are a year old, plant them out, and train them as above.

Planting and Culture.

As we before observed, may either plant large cuttings at once where they are to remain, or young trees previously raised,

raised, as above in the nursery, from four or five, to six feet high.

Plant the standards fifteen or twenty feet distance; suffer them to branch out above, and form full and spreading heads, according to their natural growth.

The hedge plants, designed to form a full hedge, chiefly for the berries, may be arranged about a yard asunder; and, in their future growth, only trim up the lower stragglers on the sides, and let them branch out freely above.

The FIG-TREE.

THE Fig-tree *(Ficus)* is famed for its singularly rich and delicious fruit: though, on account of its peculiar luscious flavour, it is not so generally palated as many other sorts; however, being an eminently fine fruit, the trees deserve culture in every good garden, principally as wall trees, and occasionally as espaliers and standards; of which there are many choice varieties, originating from one main species, belonging to the class and order *Polygamia Polyoecia*, i. e. flowers being of different sexes, on the same, and on different plants.

There is only one species of the cultivated Fig, comprising several varieties; *viz.*

Ficus Carica, i.e. CARICA, or COMMON FIG-TREE. — *Having large palmated or hand-shaped leaves,* and numerous minute flowers, concealed within a general cup or cover, becoming the fruit, pear or top-shaped, ripening in Autumn, to different colours, in the varieties; *viz.*

EARLY LONG BLUE or PURPLE FIG. *Beginning of August.*
LARGE BLUE FIG. *August and September.*
LARGE BROWN or CHESNUT FIG. *Beginning of August.*
EARLY WHITE FIG. *July and August.*

BLACK ISCHIA FIG. *Middle of August.*

SMALL BROWN ISCHIA FIG. *August and September.*

GREEN ISCHIA FIG. *End of August.*

LARGE WHITE GENOA FIG. *August.*

BROWN MALTA FIG. *August and September.*

BLACK GENOA FIG. *August.*

LONG BROWN NAPLES FIG. *September.*

ROUND BROWN NAPLES FIG. *End of August.*

BROWN MADONNA or BRUNSWIC FIG. *August and September.*

The Fig-tree grows fifteen or twenty feet high, making strong succulent green shoots, garnished with large leaves, divided more or less into five lobes;

lobes; producing the flowers and fruit always on the young wood of the former year's growth, arising in Spring, immediately from the eyes of the shoots, like small buds, each forming a sort of general cup to numerous small florets or flowers within, gradually encreasing in size till August and September, then ripening with a soft, tender, delicious pulp; and as the trees bear only on the young year old shoots, a general annual supply thereof must be retained in every part, as succession bearers, and which, as they bear mostly towards the upper parts, must not be shortened.

The fruit buds arise, both in the Spring, on the former year's wood, and in Summer and Autumn on shoots of the

the year; but it is from the Spring production we are to expect the main crops; for although the Summer and Autumn Figs attain perfection as secondary crops the same year, in warm countries abroad, they do not attain maturity in England in the open air: so they should be generally rubbed off as useless the beginning of Winter.

The trees succeed in any common soil of a garden, but their tender shoots are liable to be killed in severe Winters, in open exposures:

They should therefore, generally be cultivated principally as wall trees, in a sheltered sunny situation, both to defend the tender shoots in Winter, and promote

promote the ripening of the fruit in greater perfection; allotting the principal part for south walls; some also on west and east walls, for succession crops, planted twenty feet distance, and their branches arranged horizontally, six or eight inches asunder, retaining a full supply of each year's shoots, as before noticed, for the main bearers, trained always at full length.

They also often succeed well in espaliers, in a sunny exposure, and bear plentiful crops, managed as those against walls.

Likewise as half and full standards, planted in a sheltered sunny situation; in favourable warm dry seasons they often

produce

produce tolerable crops of very good Figs.

Method of Propagation &c.

The Fig is propagated by suckers, layers, and cuttings.

By Suckers.—Many suckers arise from the root, which, in Autumn or Spring, dig up and plant, either the strongest at once, when they are to remain, especially as wall and espalier trees, or in nursery rows for training; observing, in either case, to train them as required; if as dwarfs, for walls, &c. head them in Spring to eight or ten inches, to obtain lateral branches; and for standards, train them with stems from three

three to six feet high, then top them, and let them branch out into full heads.

By Layers.—Chuse the lower pliable young branches and shoots in Autumn or Spring, and lay them in the ground five or six inches deep, with the tops out; they will be rooted, and fit to plant off next Autumn: managing them as the Suckers.

By Cuttings.—Cut off a quantity of the young robust shoots in Autumn or Spring, from ten to fifteen inches long, and plant them with their tops entire, in a shady border, in rows two feet asunder, trained as the Suckers.

Final Planting.

As the Fig-Tree generally succeeds best when planted out finally while young, may plant as wall and espalier trees, either the Suckers, Layers, or Cuttings, as soon as rooted, at once where they are to remain, or such as have been previously trained, two or three years, and formed a head of branches; but for standards, should chuse for final planting such as are trained with tall erect stems a proper height, and have branched out at top, and formed heads.

Plant them in Autumn or Spring, the wall and espalier-trees twenty feet distance

distance at least, for the branches, being trained horizontally, will soon fill that space; and the standard trees should also be arranged twenty or thirty feet asunder, especially the full standards.

General Culture.

Then as to culture of pruning and training, observe as below.

First, of the wall and espalier trees; these being furnished with a head of lateral branches, obtained as directed in their nursery culture, let all the branches be trained horizontally to the wall, &c. ranging equally to both sides at full length, six or seven inches asunder, continuing to encrease the the number of branches annually upward,

ward, arranged at the above distance; and, if wood is wanted, may either prune short some adjacent shoot in Spring, or pinch shoots of the year, in May or June, to force out laterals, but otherwise generally train the whole supply of bearers always at full length.

Every Summer, about June, July, and August, go over the trees, both wall and espaliers, and cut out only directly fore-right, and other very irregular shoots of the year, or such as appear absolutely useless, or cannot be trained in, carefully reserving all the regular side shoots, and tacking them in strait and close, at all their length, to admit the sun and air to improve the fruit; leaving an abundant supply to chuse from in Winter pruning for next year's bearers,

bearers, not shortening any during their Summer's growth.

The Winter, or general pruning, may be performed either in November or Spring; though as the shoots of Fig-trees are very liable to suffer by severe frosts, if left unpruned till February or March, there will be a greater chance out of the whole, to have a sufficiency survive the rigours of the Winter, to chuse from in Spring pruning.

Then, in the operation of Winter pruning, must retrench old naked wood, and retain a full supply of young shoots for next year's bearers, observing, where old naked branches, unfurnished

with young wood, advance a considerable length, to prune them out either wholly, or down to some lateral shoots, &c. to supply the place: selecting a general reserve of the best-placed shoots, not only collaterals or side shoots, arising on all the main branches at eligible distances, advancing in progressive order, one after another, between the mother horizontals, from the very bottom to the extremities, but also a terminating one to each branch, and as the foremost branches advance too long, the laterals come up to supply their places; retaining generally the most robust short-jointed shoots, rejecting very long, weak, and ill-ripened ones; cutting out, as you go on, all the superabundant and useless young growths quite close, together with
all

all dead wood, and part of any too long-advanced older branches, cut down to a lower shoot; preserving the whole supply of bearers at full length; then directly nail and tye them regularly to the wall and espalier, strait and close at the aforementioned distances.

As to the Standard Figs, let them branch out freely at top, and only cut out in Spring any very irregular growths, and the ends of dead shoots, leaving all the others entire, permitting the whole to branch out, and form a full head.

The GOOSEBERRY-BUSH.

THE Gooseberry (*Grossularia*) is ranged by the Botanists as a specie of *Ribes* or Currant, according to the agreement of their fructification or parts of the flowers, and are also of the Class and order, *Pentandria Monogynia*, flowers having five stamina and one style.

There are several species of Gooseberries, comprising many varieties; and are distinguished by the Botanists as follow:

Ribes Grossularia, i. e. *Grossularia*, or COMMON GOOSEBERRY. *Having prickly erect*

erect *branches*, and *hairy berries*.

Ribes reclinatum. RECLINATED GOOSEBERRY-BUSH. *Having somewhat* reclining *branches, but moderately armed with prickles.*

Ribes Uva-Crispa.—or (UVA-CRISPA) SMOOTH-FRUITED-GOOSEBERRY BUSH, *Having erect* prickly branches, *and smooth berries.*

Ribes Oxyacanthoides — HAW-LEAVED GOOSEBERRY-BUSH. Having very prickly branches, and large leaves.

Of the above species there are many different **varieties of** the fruit; some are round, others **oval, and of** different colours, as Red, Green, Yellow, White; some **hairy,** and some smooth; generally known **by the** following names:

Y *Red*

Red Kinds.

SMALL EARLY BLACK-RED GOOSEBERRY.
HAIRY RED GOOSEBERRY.
SMOOTH RED GOOSEBERRY.
DAMSON, or DARK RED GOOSEBERRY.
RED RASPBERRY GOOSEBERRY.
GREAT MOGUL TAWNEY GOOSEBERRY.

Green Kinds..

EARLY GREEN HAIRY GOOSEBERRY.
SMOOTH GREEN GOOSEBERRY.
GREEN GASCOIGNE GOOSEBERRY.
GREEN RASPBERRY GOOSEBERRY.

Yellow Kinds.

EARLY AMBER GOOSEBERRY.
GREAT OVAL YELLOW GOOSEBERRY.

GREAT

GREAT ROUND AMBER GOOSEBERRY.
HAIRY AMBER GOOSEBERRY.

White Kinds.

COMMON WHITE or CRYSTAL GOOSEBERY.
GREAT CRYSTAL GOOSEBERRY.
WHITE VEINED GOOSEBERRY.

Different Colours.

GREAT RUMBULLION GOOSEBERRY.
LARGE IRONMONGER GOOSEBERRY.
SMOOTH IRONMONGER GOOSEBERRY.
GREAT HAIRY GLOBE GOOSEBERRY.

The bushes grow six or seven feet high, branching out low into very bushy heads, armed more or less with thorns, and

and have trilobated leaves, and small greenish flowers at the sides of the branches, succeeded by the berries, attaining a useful state in May or June, to use green for culinary purposes, and ripe for eating in June, July, and August.

They bear both on the young and old wood, immediately from the eyes of the young shoots, and on small spurs, arising on the two, or three, and many year old branches.

As the Gooseberry is a very useful Summer fruit, both when green for many culinary purposes, and when ripe very palatable and wholesome to eat, every good garden should be well furnished with plenty of the bushes, cultivated generally as bushy standards, in the kitchen garden, &c. both in single rows, and in full plantations, from six to eight or ten feet distance in each row; trained commonly with single stems a foot or more high, branching out above with regular heads, which must be kept in order by cutting out ill-placed and crowding shoots, retaining the general regular branches, six or eight inches asunder; formed either with the heads full in the middle, or concave, or hollow, by retrenching the central shoots, as observed in the Currants; leaving the shoots mostly entire, or but moderately shortened; at least till arrived to the intended height. *See their Training and Culture.*

Some may also be trained with fanned heads, espalier ways, and also to plant against walls, &c. for early fruit.

Propagation and Training.

Gooseberries are expeditiously raised in abundance from suckers and cuttings, and occasionally by layers, to a bearing state in two years.

By SUCKERS.——Abundance arise annually from the roots of old trees, which may be dug up in Winter or Spring. Chuse the strongest, and prune the crooked and weak tops a little, then plant them in nursery rows for training; or some of the strongest at once where they are to remain; and in their future growth trim off all lower shoots from

from the stem, and irregular ones in the head, keeping the general branches six or eight inches asunder, as before observed, only shortening long stragglers, &c. and they will bear the second year.

By Cuttings.—Chuse strong, straight young shoots, of the upper branches, cut off a foot or more long; plant them in rows, in a shady border, and they will readily grow; training them as the suckers.

By Layers.—Lay the lower branches almost at any time, and any how, in the earth; they will root freely, and be fit to plant off in Autumn after; training them as above.

Final Planting and Culture.

Gooseberry bushes, when advanced two or three feet in height, with tolerable

ble bushy heads, may be planted out finally to remain, any time from October till March.

Plant the main supply in the large quarters of the kitchen garden; some arranged in a single row, around the verges or borders, eight or ten feet asunder, others in wide cross rows, to divide the ground, and some in continued plantations, six or eight feet in the lines, with an interval of ten feet between the rows, as advised for the Currants.

May likewise plant a few fanned trees against warm walls, to produce some early fruit.

Then, as to future management, it is nearly

nearly the same as directed for Currants, observing also as below.

Keep the bushes always to one single stem below, by clearing off all lower shoots, and suckers from the root. The head should be kept open, and regular, and the general branches continued about six or seven inches asunder, to have large fruit; and shortened but little, or hardly at all, in pruning, except very long ramblers, or low stragglers; retain the same branches as bearers, as long as they support a good bearing state, because those of several years old bear abundantly on spurs; but according as they gradually become very old or worn out, and produce but small fruit, should retain strong young shoots, advancing below, to supply their place, according to the following rules of general pruning.

As these bushes generally emit numerous shoots every Summer, you may, if any are very crowded, in June or July, prune out close some of the most irregular and crowding, to open the head, for the more free admission of the sun and air, to promote the size and goodness of the fruit; retaining enough of the regular shoots entire to chuse from in Winter pruning, if required.—And in Winter pruning, proceed to prune out close all the superabundant lateral shoots of last Summer, together with any very crowded, irregular, older wood, and old worn-out bearers; retaining young ones advancing below in their room, and leaving the general branches about six or eight inches distance, at least, as before observed; each terminating in a young shoot;

shoot, either naturally; or if any branches are advanced too long or straggling, prune them down to a lower shoot for a leader; preserving all the small lateral fruit spurs; and let the occasional supply of young shoots be but moderately shortened; principally only long rambling growths, and very bending and straggling shoots, just reduced to a little regularity, as observed for the Currants.—The bushes will thus shoot moderately, and produce larger and finer fruit in proportion.

For if Gooseberries are too generally shortened, it forces out numerous unnecessary, useless shoots, from all the lower eyes, in Summer, forming the head a perfect thicket, rendering the fruit small, and occasioning much trouble of pruning in Winter.

Never prune them with garden shears, as sometimes ignorantly practised; but always with a knife: Likewise, always keep the branches of the head thin, and you will have large handsome berries, ripening with a good flavour.

FIL-

The HAZEL and FILBERT-NUT-TREE, &c.

THE Hazel *(Corylus)* comprises the common Wood Nut, the Filbert, Barcelona, Cob Nuts, &c. all varieties of one species, which merit culture in the fruit collection, for the variety of their fruit; and belong to the class and order *Monoecia Polyandria*, flowers male and female on the same tree; the males having numerous stamina.

The Hazel consists of one species, comprising several varieties.

Corylus Avellana, i. e. *Avellana*, or COMMON HAZEL-NUT-TREE, having round leaves, with *oval obtuse stipula*, at the

the base; male flowers in long catkins, and female flowers close to the branches, succeeded by the nuts, in large torn cups, consisting of the following varieties, ripe in Autumn; viz.

Common Wood Nut, with white, and with red skinned kernels.
Large cluster Wood Nut.
Filbert Nut, with red kernels.
Filbert Nut, with white kernels.
Barcelona Nut.
Cob Nut, (very large).
Long Nut, (a curiosity).

The Hazel tree, and varieties, grow from about twelve or fifteen, to twenty feet high; very branchy, almost to the bottom, having roundish rough leaves, and

and produces the flowers and fruit from the sides of the branches, being male and female, flowers separate, without petals; the males in long yellowish amentums, in Winter, and the females in close-fitting clusters, in the Spring, succeeded by the clusters of Nuts, inclosed each in its torn calix or cup, ripening in August and September.

The Common Wood Nut Trees grow in vast abundance in our woods and hedges, and are sometimes admitted in Gardens, &c. for variety.

But the other larger sorts, being improved varieties, are cultivated more abundantly in gardens and orchards, but the Filbert most of all, for the goodness of of its kernel; though the large Cluster and Barcelona Nuts are also well worth cultivating, and the Cob and Long Nut as singular varieties: but where there is good scope of ground it is worth attention to cultivate some of each sort in gardens, orchards, and fields, &c. as they will succeed almost any where.

They may be employed both as full and half standards; planted either in continued ranges, fifteen or twenty feet asunder, or dropped singly in different parts to effect variety, keeping them to clean stems, and let the heads branch in their natural manner.

Some may likewise be arranged hedgeways,

ways, five feet distant in the row, and suffered to take their natural growth.

And, for variety, some may be planted in espaliers, especially the Filberts, arranged fifteen feet asunder, with the branches trained at full length, retaining occasional lateral shoots, advancing below as succession bearers. They are all very hardy, and will succeed in any common soil and exposure.

Method of Propagation and Training.

All the varieties of the Hazel-tree are propagated both by planting the Nuts, and by suckers, layers, and grafting; but it is proper to remark, that as the seedling-raised plants are apt to vary, the latter methods are the most certain, whereby to continue the improved varieties distinct.

By the Nuts.——Let a quantity of the best Nuts of the desired sorts be preserved in sand till February, then sowed in drills two inches deep, and when the plants are a year or two old, plant them out in nursery lines, training some as full, half, and dwarf standards, with six, four, and two feet stems; then top them, and permit them to branch out above, and form heads; some also as hedge plants, branching out near the bottom, but fanned up on the sides.

By Suckers.—The trees send up abundance from the root; which being dug

dug up in Winter, or Spring, may either plant the strongest at once, to remain, especially if for hedge-rows, or wholly in the nursery way, to train for the purposes intended.

By LAYERS.—Any time from Autumn till Spring, in open weather, have recourse to some low branching trees, chusing some of the lower pliable branches, furnished with plenty of young shoots: peg them down, and lay all the young wood three or four inches deep, with the tops out, which by next Autumn will be rooted; then plant them out, and train them as the seedlings.

By GRAFTING.—This may be practised to continue and improve any particular

cular fine variety, by inserting shoots thereof into any kind of Hazel stocks, in the common method of grafting.

Planting and Culture.

When the trees are advanced from about four to five or six feet high, they may be planted out to remain; arranging the standards, some in full plantations, others dropped singly in borders, shrubbery clumps, or on the sides of shady walks, &c. and others in hedgerows, as aforesaid.

In Culture, keep the standards to clean single stems, and let the heads branch out in their own way, only reforming casual long ramblers, or low stragglers;--And in the hedge plants, trim up only low straggling branches, and let them branch out above at sides and top.

The MEDLAR TREE.

THE Medlar Tree (*Mespilus*) is singular in its production of a large apple and pear-shaped brown fruit, not eatable until it assumes a state of decay; for while it remains firm and sound it is exceedingly austere; but ripening in Autumn, becomes soft and palatable in Winter; the tree belongs to the class and order *Icosandria pentandria*: Twenty or more stamina, and five styles, in each flower.

There is but one species of the common Medlar, which, by the Botanists, is called

Mespilus Germanica,—GERMAN, or Common

mon *Medlar Tree*; described *Medlar, with smooth or thornless branches, spear-shaped leaves, downy underneath, and large solitary, or single-placed, close-sitting flowers,* succeeded by large, round, umbilicated, brown, hard fruit, the size of middling apples, each containing five stoney seeds, ripening in October and November, but not eatable till after being gathered and lain some time, when it becomes soft, tender, and palatable, of a sharp, vinous flavour, and comprises the following principal varieties, viz.

COMMON LARGE GERMAN MEDLAR.
NOTTINGHAM, or SMALLER MEDLAR.
PEAR-SHAPED ITALIAN MEDLAR.

The Medlar-Tree grows deformedly, twelve or fifteen feet high, or more, ornamented

namented with large lanceolate leaves; and produces its flowers and fruit fingly, from the sides of the two or three, and many year old branches, often upon short spurs; the same branches continue bearing many years, and should generally be permitted to run without shortening, except reducing casual stragglers, &c.

This tree certainly merits admittance in every collection, for the singularity and variety of its fruit; trained principally as standards and half standards, in the garden and orchard, or in any out grounds, for they will succeed any where; and some also, occasionally, as Espaliers, in assemblage with quinces, apples, &c. managed as for apples and pears.

Method

Method of Propagation and Training.

The trees may be raised from seed; but by grafting and budding may preserve the varieties permanent, and raise them sooner to a bearing state. Also by layers.

By SEED.———Sow the kernels of the fruit in a bed of light earth, an inch or more deep, in November or December; they will, probably, come up in Spring following, and, when a year or two old, plant them out in the nursery way, and train them with from four to five or six feet stems, for half and full standards; then let them branch out at top, and form heads; but those designed for espaliers should be topped in their infancy to ten or twelve inches,

inches, to provide lower branches, as directed for Apples, &c.

By GRAFTING, &c.—The improved varieties may be grafted or budded, either upon Medlar, Pear, Quince, or White-thorn Stocks, raised as above, for the seedling Medlars, several feet high for standards, and lower in proportion for dwarfs; then let them be grafted or budded, and trained for the purposes intended, as directed for the Apple and Pear tree.

By LAYERS.—In Autumn, Winter, or Spring, lay the lower young branches in the common way, which will be rooted by Autumn following; then plant them off, and train them as the seedlings.

Planting and Culture.

When raised to a due size, plant the standards any where in the garden, orchard, shrubbery, &c. twenty feet asunder, and suffer them generally to grow in their own way, only pruning occasionally any very irregular, rambling, and straggling growths, as advised for other standard fruit trees.

Plant the espaliers fifteen or eighteen feet asunder, training the branches horizontally, five or six inches distance, mostly at full length, to their allotted extent, except in their first training; when, if required, may prune short occasional young shoots in Spring or Summer, to encrease the number of branches to cover the espalier regularly; observing always to give

give them a Summer and Winters pruning and training, exactly as directed for Apples and Pears.

Of the Fruit.

The Medlars attain full growth on the trees the end of October, or beginning of November, at which time let them be gathered, and prepared to an edible state, by laying some in moist bran, others in dry straw, and some on the shelves of the fruitery, &c. those in the bran will become soft and eatable in two or three weeks, and the others will succeed them.

The MULBERRY TREE.

THE Mulberry Tree (*Morus*) is in high eſtimation for its fine juicy berries, very refreſhing to eat in the hot weather of Autumn, as well as for ſome culinary uſes, and makes a fine ſyrup; it is likewiſe eminent for its leaves to feed ſilk-worms, and belongs to the claſs and order *Monoecia tetrandria*, i. e. male and female flowers apart on the ſame tree, the males having four ſtamina.

There are ſeveral ſpecies of Mulberries; but there is only one ſort commonly cultivated as a fruit tree in England; viz.

Morus

Morus Nigra, Black or Common Mulberry Tree, *having large heart-shaped rough leaves*, and small male and female flowers, without petals; the males growing in amentums or strings, and the females in roundish close heads, which succeed to an ovalish, very succulent, blackish red berry, full of tuberances, each having one seed, and ripens in August and September; comprehending two varieties, viz.

Common Black Mulberry.
Jagged-leaved Smaller Mulberry.

The following species are also sometimes cultivated for variety.

Morus Alba, White Mulberry.
Morus Rubra, Red Mulberry.

But

But the Common Black Mulberry is the principal sort to cultivate for the general supply.

The Mulberry tree grows twenty feet high, or more, dividing low into a large branchy, widely-spreading, full head; it produces its flowers and fruit on the young wood; very short robust shoots of the same year arising from the ends of others produced the Summer before, towards the upper parts of the year-old branches; in which case they must generally be permitted to remain entire, as shortening would cut away the fruitful parts, and retard their bearing.

The trees are very hardy, and succeed in any common soil, trained both as common

mon standards, half standards, and dwarfs; and likewise as wall and espalier trees.

But common standards is the most general mode of cultivating Mulberry trees, planted detached, either in the garden, orchard, or on lawns, or on any plat of grass ground, in a warm, sunny exposition; though a few trees may be sufficient for private use, as sometimes one or two large ones furnish berries enough to supply a numerous family.

May also cultivate some both as wall and espalier trees, for variety, and they will produce larger, earlier, and richer flavoured berries.

Method

Method of Propagation and Training.

The Mulberry tree is propagated by layers, cuttings, grafting, and occasionally by seed.

By LAYERS.—Chuse the young branches in Autumn or Spring; but unless they are situated low, or that there are stools previously formed, by heading down some young trees near the ground, to furnish bottom shoots, commodious for laying, we must use large pots, boxes, or baskets of earth, raised upon stands near the branches; laying the young shoots three inches deep, with their tops out; by next Autumn, being rooted, plant them off in nursery-lines, and train them
with

with clean stems, five or six feet high, for standards, or in proportion for half standards, &c. but when designed as dwarfs for walls or espaliers, they should be headed while young to ten or twelve inches, to obtain lower branches, as formerly observed for other dwarf trees.

By CUTTINGS.—In Autumn or Spring cut off a quantity of last year's strong shoots, from eight or ten, to twelve inches long, retaining their tops entire, and plant them in a shady border, to remain till Autumn or Spring following; then, the plants being rooted, set them out in the nursery, to be trained like the layer-raised plants.

By GRAFTING.—This is sometimes practised to continue and improve any desirable variety, by inserting shoots thereof into stocks of the Mulberry kind.

By Seed.—Sow them in March, on a warm border, half an inch deep; and when the plants are a year or two old, bed them out in rows, and train them as above.

Planting and Culture.

Plant the standard Mulberries in any dry, sheltered, sunny exposure, either together, thirty feet distance, or detached in different places; and let them generally take their natural growth, retrenching casual dead wood, or any very ill growing or low straggling branch; but never shorten the shoots, and they will then bear plentiful crops of berries annually.

Plant the wall and espalier Mulberries,

ries, fifteen or twenty feet diſtance, and arrange their branches horizontally, five or ſix inches aſunder, at full length, while the trees are young, only ſhortening occaſional middle ſhoots in Spring or Summer, to force out a farther ſupply of wood, to furniſh the wall and eſpalier more effectually; encourage alſo a ſucceſſion of regular ſide ſhoots, gradually advancing in every part, from the bottom upward, as bearers; and as any very old, long-advanced, naked branches appear, prune them down in Winter, and introduce lower, advancing young wood in their ſtead; training the whole ſtraight and cloſe to the wall and eſpalier, at full length, becauſe ſhortening the ends cuts away the only fruitful parts, as before remarked.

The NECTARINE TREE.

THE Nectarine tree (*Nuciperfica*) is of the family of *Amygdalus*, or Almond, and by the Botanists is considered generally as a variety of the Peach, as some have affirmed to have seen Peaches and Nectarines growing naturally on the same tree; and is of the class and order *Icosandria monogynia*, twenty or more stamina, and one style, within the same flower.

The modern Botanists give the Nectarine no specific distinction, only as a variety of the Peach, by the old name of *Nuciperfica*, the NECTARINE TREE, which, like the Peach, has *spear-shaped, acutely-sawed*

sawed leaves, without glands at the base, which obtain in the Almond *close-sitting, solitary, or singly-placed flowers*, having *five red-ish petals*, succeeded by a large, round, firm, smooth fruit, having a nut or stone in the centre; and consists of several valuable varieties, ripening in August and September to a reddish and yellow colour; viz.

FAIRCHILD's EARLY NECTARINE. *Middle of August.*

ELROUGE NECTARINE. *End of August.*

VIOLET NECTARINE. *Middle of August.*

NEWINGTON NECTARINE. *Beginning of September.*

TEMPLE NECTARINE. *Middle of September.*

RED ROMAN NECTARINE. *Middle of September.*

MURRY

MURRY NECTARINE. *Beginning and middle of September.*

BRUGNON, or ITALIAN NECTARINE. *September.*

SCARLET NECTARINE. *Beginning and Middle of September.*

TAWNEY NECTARINE. *September.*

GOLDEN NECTARINE. *Middle or End of September.*

VIRMASH, or PETERBOROUGH NECTARINE. *October.*

All the varieties of Nectarines have a smooth rind, and firm rich pulp, which in some adheres to the stone, and in others it separates.

The Nectarine tree grows in every respect like the Peach, not distinguishable one from

from the other but by the fruit; it rises fifteen feet high or more, making long green shoots, garnished with long, narrow, spear-shaped leaves, and produces its flowers and fruit also on the young shoots of a year old, immediately from the eyes of the shoots; and, as the same shoots generally bear but one year, except on casual spurs, a general supply of new ones of each year is requisite in every part of the tree, annually, as succession bearers, and which, in their training as wall trees, should be retained at full length all Summer, and commonly pruned more or less in Winter, to promote a more certain succession of new bearers from the lower eyes, as explained for Apricots and Peaches.

But

But although the Nectarine and Peach trees discover not the least difference in their general growth, yet there is a very obvious distinction in their fruit; the Nectarine having a smooth hard rind, while that of the Peach is downy and soft.

As the Nectarine tree blooms early in Spring, previous to the expansion of the leaves, and before the weather sets in warm, and, the young embryo fruit being tender, liable to suffer by the severity of the cold, should therefore, like the Apricot and Peach, be allowed a warm, sunny, exposure, against South, and West, and East walls, &c. trained both as common dwarf wall trees, for the general supply, and as half standards between the dwarfs, as observed for Apricots, &c.

It

It is a hardy tree in respect to growth, and succeeds in any good garden earth.

Method of Propagation and Training.

The Nectarine is propagated always by budding the approved varieties on Plum stocks, and occasionally on Almond, Peach, and Apricots; but principally on plums, as being the hardiest and most successful stocks for the general supply, raised from the stones of the fruit, as directed in the culture of Almonds and Apricots, to a proper size to bud for dwarfs and half-standards, performing the budding also as directed for those trees, Peaches, &c.

The buds shooting forth in Spring following, and forming each one strong shoot,

shoot, three or four feet long, by the end of Summer, the trees will then have formed their first head as Nectarine trees, proper either for final transplantation into the garden, against South walls, and some on West and East expofures, as before obferved, arranged fifteen or twenty feet diftance; or may remain longer in the nurfery, trained in a fanned manner againft any kind of clofe fence, reed hedges, or to ftakes, &c. till wanted.

But obferving the firft head muft be cut down in March to fix or eight inches, to force out four or more laterals below, in Summer, to give the tree its firft fpreading form for the wall, training the fhoots horizontally to the wall &c. at full length, all Summer, till Winter or Spring; then prune each to eight

eight or ten inches, to gain a farther supply of branches, and thus continue encreasing their number annually, by shortening the shoots more or less, both in Winter or Spring pruning, and by pinching shoots of the year in Summer, as observed for Apricots and Peaches.

But as to the general culture of the full trained Nectarine trees, you must, as in Peaches, preserve a general supply of the best placed young wood of each Summer as bearers, the future year, training them at full length all Summer, cutting out all foreright and luxuriant wood; and in Winter selecting plenty of the best-placed, last Summer shoots, in every part, quite from the bottom upward, at regular distances cut out the irregular and superabundant

dant ones, together with part of the old bearers, down to the young wood, and shorten the supply of new shoots from six or eight, to fifteen or eighteen inches, according to their strength; then nail the whole regularly to the wall, five or six inches distance.

Observe exactly the same culture as for Peaches.

The PEACH TREE.

THE Peach tree, (*Persica*) a species of the *Amygdalus*, or Almond, produces the most valuable of our stone fruits, surpassing most others in their vast variety, beauty, richness of flavour, and long duration in perfection, they continuing in eating from July till November, in the different varieties, wherey the trees demand our principal care, as the choicest of wall trees, planted against the best southerly exposures; and belongs to the class and order *Icosandria monogynia*, as the Almond.

There is but one species of the Peach tree, and which being considered by the Botanists

Botanists as a species of the Almond, (*Amygdalus*) they call it,

Amygdalus Persica,—PERSICA, or the PEACH TREE, *having spear-shaped, sharply-sawed leaves, without glands* at the base, (as in the Almond) and solitary, or singly-placed flowers of five petals, succeeded by a large, round, downy fruit, consisting of a delicious, juicy pulp, with a nut or stone in the centre, containing one kernel; and comprises many eminent varieties, ripening from July till October, in the different sorts; which are distinguished by the following names;

EARLY NUTMEG PEACH (very small). *July.*
EARLY ANNE PEACH (small). *Beginning or middle of August.*

SMALL

SMALL MIGNON PEACH. *Middle of August.*

EARLY PURPLE PEACH. *Toward the middle of August.*

WHITE MAGDALEN PEACH. *Middle of August.*

RED MAGDALEN PEACH. *End of August.*

GREAT MIGNON PEACH. *Middle of August.*

EARLY NEWINGTON PEACH. *August.*

OLD NEWINGTON PEACH. *Middle of September.*

CHANCELLOR PEACH. *End of August.*

LATE PURPLE PEACH. *Middle of September.*

MONTAUBAN PEACH. *End of August.*

NOBLESSE PEACH. *Beginning of September.*

BELLE GARDE PEACH. *Beginning and middle of September.*

BELLE CHEVREUSE PEACH. *August.*

YEL-

YELLOW ALBERGE PEACH. *Middle and end of August.*

LA TETON DE VENUS PEACH. *Middle and end of September.*

PERISQUE PEACH. *End of September, and early in October.*

RAMBOUILLET PEACH. *End of September.*

BOURDINE PEACH. *Beginning and middle of September.*

NIVETTE PEACH. *Middle and end of September.*

ROSSANA PEACH. *Beginning and middle of September.*

SION PEACH. *End of September.*

ADMIRABLE PEACH. *Middle and end of September.*

ROYAL PEACH. *End of September.*

VIOLET PEACH. *Middle of September.*

PORTUGAL PEACH.

ROYAL

ROYAL GEORGE PEACH. *Early in September.*

CATHARINE PEACH, (very large and beautiful). *October.*

BLOODY PEACH. *Middle and end of October.*

CAMBRAY PEACH. *October.*

NARBONNE PEACH. *October.*

MONSTROUS PAVIE of POMPONNE, (exceeding large). *End of October.*

The Peach tree grows fifteen feet high, or more, but is commonly trained as a wall tree; emitting numerous long, smooth green shoots, ornamented with long, spear-shaped sawed, smooth leaves; and produce their flowers and fruit along the sides of the young wood of the former Summer, immediately from the eyes of the shoots, which always produce at the same

same time, both a crop of fruit, and a supply of new shoots for next year's bearers; for the same wood rarely bears much fruit but once, except sometimes on some casual small spurs, arising on the two-years wood; but must depend always on the last year's shoots for the main crop of fruit; a general annual supply of which must be retained in every part of the wall trees, as succession bearers, at full length, all Summer, and shortened in Winter pruning, to encourage succession bearers from the lower eyes, which would otherwise arise mostly towards the extremities, and leave the bottom naked; for, as in the Apricot, the general supply of the immediate bearing shoots rise principally only from the year old wood, produced the Summer before.

Peaches

Peaches should be trained principally as wall trees, for as they blossom, and set their fruit early in Spring, in cold weather, and their young infant fruit being of a tender nature, they seldom bear well, or ripen the fruit in due perfection in the open ground, on espaliers and standards; so should plant them mostly against the best South walls, and occasionally on West and East walls, for successional fruit.

Generally have the main plantation composed principally of the common dwarf wall trees, with low stems, branching out near the ground, planted at fifteen or twenty feet distance, with the branches arranged horizontally, four or five inches asunder.

Others may be employed occasionally, as quarter, half, and full standard wall trees,

trees, formed with fanned, spreading heads, and planted between the dwarfs, against high walls, in order to occupy the upper parts, while the common dwarf trees are advancing below.

A few trees may also be planted in espaliers, and as detached half standards, in the open ground, for variety, to take their chance.

The Peach tree prospers in any good garden earth; but if of a rich loamy temperature it may prove an additional advantage.—*See Planting*.

Propagation and Training.

The Peach, in all its varieties, is propagated, and each sort continued permanent,

nent, by budding them upon Plum stocks, or occasionally on Peach, Almond, and Apricot stocks.

But for the main supply prefer chiefly plum stocks, as the most hardy and durable; raised either from Suckers arising from plum tree roots, or from the stones of the fruit, as directed for Almonds; which being planted in nursery lines, will in two years be of due size to bud in July or August, within half a foot of the bottom, for common dwarf wall trees, but for half standards, &c. must be run up three or four years, with tall stems, then budded, at from three or four, to five or six feet high; inserting generally but one bud in each stock.

Then, in Spring following, the buds will advance, each with one strong shoot, attaining three or four feet in length by next Autumn, forming the new tree, which may then, if required, be planted in the garden, finally to remain.

In March following, when the head, or first shoots are about a year old, let them be headed down to a few eyes, to gain four, five, six, or more laterals, below, in Summer, to form the head, training them horizontally to the right and left, in a fanned manner.—*See the General Culture.*

Planting and Culture.

Peach trees may be planted any time in open weather, from October till March, and

and may either plant those of only one year old, with the first head entire, or such as have been headed and trained in the nursery, or that have arrived to a bearing state, thereby furnishing your wall at once with immediate bearers; having the trees digged up with their full roots; prune off only broken ones, and reduce very long stragglers, leaving the head entire at present, then plant them eighteen or twenty feet distance, as before mentioned, and tack their heads to the wall.

Then, in regard to general culture, it consists in giving them eligible pruning and training every Summer and Winter, of which take the following observations.

The

The first pruning consists, as we before hinted, in having the first shoots, or head, produced immediately from the budding, headed down when a year old, in March, to a few eyes, if not done in the nursery, as directed in their nursery culture, to procure several lateral branches, proceeding immediately near the head of the stem, trained at full length all Summer, and in Winter or Spring following, shorten also all these second shoots, generally from about eight to ten or twelve inches, leaving the lower ones the longest; and pinch young ones of the year, in Summer, in order to gain annually a farther requisite supply of lower branches, to give the head a good spreading form, advancing regularly quite from the bottom upward, with the branches arranged horizontally, equally

to

to both sides, four or five inches distance, at full length all Summer, shortened more or less always in winter pruning, according to their strength, as hereafter exhibited.

Then, with respect to the general Summer and Winter pruning, observe as below.

Every Summer, in May, June, and July, go over the trees to regulate the young growth of the year, by rubbing or pruning off all the fore-right buds or shoots, with other ill-placed growths, and very luxuriant or rank wood, both to admit the free air, and the benefit of the sun to the fruit, as well as to disburthen the trees of unnecessary and useless shoots; retaining a plentiful supply of all the young, moderately-strong regular side shoots, for next year's bearers, clearing off all others that are too numerous or unnecessary; and train in the reserved supply of regular wood close to the wall, at full length, till Winter pruning.

And, in Winter pruning, we must keep in view a general reserve of the best well-placed, last Summer's shoots, for the ensuing year's bearers, in every part, from the bottom to the extremities, at regular distances, generally retaining one at least, on every one of the last year's horizontals, or sometimes two, or more, in wide or vacant spaces, as the case requires; chusing always the most promising and best-placed, and retrenching close all the superfluous, or too abundant ones, very

very rank, and all ill-placed shoots, and very weak twigs, with part of most of the last year's bearers, and old horizontals, pruned down to the first best shoots they support, together with any old, long, naked branches, unfurnished with young wood, cut either entirely out, or down to any good lateral shoot, or eligible young branch, furnished therewith; cut out, likewise, all dead wood and old stumps; thus clearing out all useless and bad growths, to make proper room to train the necessary supply of young bearers, which, as you go on, should generally be shortened, more or less, as directed below.

Likewise cut off all small lateral twigs, arising on the sides of the main shoots, retained for bearers.

But any small fruit spurs arising on

the two years wood, may be preserved, as they often afford good peaches.

As you proceed in this general Winter-pruning, shorten most of the reserved supply of young bearing wood, that they may produce succession-bearing shoots from the lower eyes; otherwise will furnish them only towards the top, and leave the bottom naked, being careful, however, to prune them according to their strength, the moderate ones to eight, ten, or twelve inches, the stronger ones to twelve or fifteen, and the strongest prune from about fifteen or eighteen inches, to two feet long; and very strong luxuriant shoots of a generally vigorous-shooting tree, may be left still longer in proportion, because, if vigorous shoots are cut

cut short, it promotes a still more luxuriant growth, without fruitfulness.

Observe, likewise, in shortening the bearing shoots, not to cut below all the blossom buds, but generally cut to a wood bud, or twin blossom, as explained in the Apricot.

As soon as a tree is thus Winter pruned, let it be directly nailed regularly to the wall, ranging the branches and shoots equally to both sides, strait and close, four or five inches asunder.

The PEAR TREE.

THE Pear tree *(Pyrus)* is the original species of the genus *Pyrus*, which comprehends also the Apple *(Malus)*, and the Quince *(Cydonia)*, as species of the same genus, each of which, however, is treated of separately under its proper head, and, like the apple, is remarkable for the comprehensive variety of its fruit, which may be divided into three orders, viz. Summer, Autumnal, and Winter Pears, all the offspring of one common species, belonging to the class and order *Icosandria Pentagynia*, being twenty or more stamina and five styles in flower.

There being but one real species of the Pear tree, and being the original of the genus *Pyrus* aforesaid, the botanists call it,

Pyrus communis, COMMON PEAR TREE, *having oval serrated leaves, and corymbous clusters of flowers on long peduncles*, composed of five white petals, with a germen underneath, becoming a pyramidal fruit, mostly extended at the base, and comprehend numerous varieties, ripening in successive order, from July till October, viz.

Summer Pears.

Summer Pears comprehend the early varieties, that ripen for eating immediately off the tree, attaining perfection in the different sorts, from July till September, and will not keep long; some only a few days after they are fully ripe, others not above a week or fortnight, viz.

LITTLE YELLOW MUSK PEAR. *July.*

GREEN CHISSEL PEAR. *Middle and end of July.*

CATHARINE PEAR. *End of July and beginning of August.*

RED MUSCADELLE PEAR. *End of July and beginning of August.*

JARGONELLE PEAR, (large and fine). *Middle of August.*

CUISSE MADAME, or WINDSOR PEAR, (large and beautiful). *Middle and end of August.*

EARLY RUSSELET PEAR. *Middle of August.*

GREAT

GREAT BLANQUETTE PEAR. *Middle of August.*

LITTLE BLANQUETTE PEAR. *Middle or end of August.*

MUSK ROBINE PEAR. *End of August.*

AUGUST MUSCAT PEAR. *End of August.*

ORANGE MUSK PEAR. *End of August.*

PERFUMED PEAR. *End of August.*

RED ORANGE PEAR. *End of August.*

SUMMER BON CHRETIEN. *Beginning of September.*

SUMMER BERGAMOT PEAR. *Middle of September.*

ORANGE SUMMER BERGAMOT. *End of September.*

ROSE WATER PEAR. *Middle of September.*

SALVIATI PEAR. *September.*

CRAWFORD PEAR. *September.*

GREEN MUSK PEAR. *September.*

LONG-STALKED BLANQUETTE. *September.*
PEAR PIPER. *September.*
ST. JAMES'S PEAR. *September.*
LEMON PEAR. *September.*
RED ADMIRABLE, (large). *September.*

Autumn Pears.

Comprehend a list of choice eating Pears, that attain perfection for eating from about the end of September, or beginning of October, till November and December, and are mostly of a handsome size, and some fine and large.

AUTUMN BERGAMOT. *End of September and beginning of October.*

BROWN BEURRE PEAR, (very fine). *Beginning of October.*

WHITE BEURRE. *Beginning or middle of October, till November.*

RED

RED BEURRE. *Beginning or middle of October.*

SWISS BERGAMOT. *End of September.*

GREAT RUSSELET. *End of September.*

MONSIEUR JOHN. *End of October, till December.*

SWAN'S EGG. *End of October, and November.*

CRESANE PEAR, (large and fine). *End of October till December.*

MUSCAT FLEURY PEAR. *October till December.*

ROUSSELINE PEAR. *End of October till November.*

MARQUIS'S PEAR.

AUTUMN VERTE LONGUE. *October till December.*

GREY GOOD-WIFE. *End of October, November, &c.*

BEURRE BERGAMOT. *October, November, &c.*

FRENCH BERGAMOT. *October, November, &c.*

POUND PEAR, (very large) *End of October, November, &c.*

GREEN SUGAR PEAR. *End of October and November.*

Winter Pears.

Under this head is comprised a valuable collection of the finest and richest eating Pears, with some that are eminent for baking, and other culinary purposes; attaining full growth on the trees towards the middle and end of October, but not maturity for eating, till after being gathered and laid some considerable time in in the house; some, probably, three, four, or six weeks, others two, three, or four months; ripening as they lie, in successive order, from November till Spring and Summer following, in the different varieties, as below.

Winter Beurre. *December and January, &c.*

Winter Bergamot. *November till January, &c.*

St. Germain Pear, (large and fine). *December till Spring.*

Colmar Pear. *December till January, &c.*

Vergouleuse Pear. *December and January, &c.*

Holland Bergamot. *January till April.*

Winter Russelet. *January, February, &c.*

SPANISH BON CHRETIEN. *December, January, &c.*

WINTER BON CHRETIEN. *February and March, till May.*

CHAUMONTELLE PEAR. *December till March, &c.*

WINTER VERTE LONGUE. *December, January, &c.*

MARTIN SEC PEAR. *November till January.*

DAUPHINE PEAR. *End of November, December, and January.*

MARTIN SIRE. *December and January.*

ST. MARTIAL PEAR. *January or February, till May, &c.*

GERMAN MUSCAT. *January or February, till April or May.*

WINTER THORN. *December till February, &c.*

EASTER

EASTER BERGAMOT. *January or February, till April or May.*

GOOD LEWIS. *December, &c.*

AMBRETTE PEAR. *December and January.*

EASTER ST. GERMAIN. *February till April, &c.*

ST. AUSTIN. *December till February.*

The following are very large, hard Pears, in estimation chiefly for baking, and other culinary purposes.

UNION, or UVEDALE'S ST. GERMAIN. *November till May.*

BLACK PEAR OF WORCESTER, (remarkable large) *November till March or April.*

CADILLAC. *December till April or May.*

DOUBLE FLEUR. *January till May.*

The Pear tree grows thirty feet high, or

or more, branching in some sorts erectly, in others more spreading and extended; and produces its flowers and fruit upon spurs or studs, emitted numerously from the sides and ends of the branches, of from three or four, to many years old; the same branches and spurs continuing improving in fruitfulness many years; the branches encreasing the number of fruit spurs annually, as they advance in length; arising first towards the extreme parts, then, by degrees, laterally, all along the sides; which shews, that the the branches must not be shortened in their general growth, because it would both cut away the first fruitful parts and force out immense quantities of useless lateral wood shoots, instead of fruit spurs, and retard their bearing.

All the varieties of Pear trees are cultivated succefsfully, both as common high ftandards, and half ftandards; and as wall and efpalier trees for the choicer forts.

But may plant a larger fhare of common detached ftandards, for the main fupply, in the open quarters of the garden, orchard, &c. arranged thirty or forty feet diftance; the heads permitted to branch out freely to their full extent, without fhortening the branches, and they will naturally furnifh bearing fpurs abundantly.

Should allot, principally, the prime forts of Pears for walls and efpaliers, in order to forward and improve the fize, beauty,

beauty, and flavour of their respective fruits; for example:

Have a few trees of the best Summer Pears; a more plentiful supply of the finest Autumn kinds; together with a principal share of the choicest Winter varieties; plant some against South-East, or West, or East walls, and others in well-exposed, sunny espaliers; or some of the Summer, and forward Autumn kinds, on North walls; but allow the Winter Pears principally a good sunny exposure; arranging the whole twenty feet distance, at least; but if twenty-five or thirty feet, they will in the end prove abundantly more fruitful, by having full scope to extend the branches horizontally, six inches asunder, without shortening,

ening, and they will emit fruit spurs along their sides the whole length.

May also train some as half and full standard wall trees, formed with tall stems, and fanned heads, to plant against high walls or buildings, or between the common dwarf wall trees.

However, in default of sufficient wall and espalier room, may cultivate any of the sorts as common standard trees, in the open ground. All the Summer and Autumn Pears will ripen abundantly well; and most of the Winter kinds also often attain tolerable perfection on standards.

The Pear-tree is very hardy, succeeds in any good fertile soil capable of yielding eligible crops of kitchen herbage, grafs, or corn, &c. as remarked of other fruit trees.

Propagation and Training.

All the varieties of Pears are propagated and continued diftinct by grafting, and budding them upon any kind of feedling Pear ftocks, and occafionally on Quinces, to form more dwarf or moderate growers; for low walls and efpaliers, &c. may chiefly rife ftocks of any of the Pear kind for the general fupply, raifed and planted in nurfery rows, as directed for Apples, to be trained to a proper fize to graft or bud, at fix feet high for full, and lower in proportion for half and quarter ftandards, and within a foot of the bottom for dwarf, wall, and efpalier trees; inferting generally but one graft, &c. in each ftock;

stock; and when they have shot, and formed their first heads of one Summer's growth, they may either be planted out in Autumn, &c. to remain, if required, or may be previously trained in the nursery.

Observing, the standards may either have the first shoots only, cut over in March to a few eyes, to obtain laterals for forming a more full and low-spreading head, or permitted to run and form a more erect and lofty handsome growth.

But the wall and espalier young trees should generally have the first head pruned down low, at a year old, in Spring, to gain a fuller supply of lower horizontals, regularly from the bottom ; and, if necessary, some of the middle-most of these second shoots may also either be pinched short

short the same year, early in June, or pruned down in Spring following, to promote more speedily and effectually a farther supply of branches to furnish the wall, &c. regularly upward. Observing to rub off early in Summer all foreright shoots, and train in the regular side ones horizontally, at full length, five or six inches asunder: continue also, after this, to train most of the future requisite supply of horizontals always entire, at the same distance, to form bearing branches, which in those trees must not be shortened.

Thus the young Pear trees, raised and trained as above, continuing their general branches entire, they, both standards and wall trees, &c. will begin to emit studs or spurs for bearing, when from two or three to four or five years old.

Planting

Planting and General Culture.

Pear trees, of from one or two to five or six years old, having formed heads, are proper for final transplantation, in Autumn, Winter, or Spring; planting them in the common method; [*See Planting.*] the standards thirty or forty feet distance, as aforesaid, and the wall and espaliers not less than twenty; supporting the tall standards erect with stakes till rooted afresh, and training the branches of the dwarfs horizontally to the wall and espalier.

Then, as to future culture of pruning, &c. the standards require but little; but the wall and espalier trees require it annually, in Summer and Winter.

Let the standards extend freely above; only prune out any very irregular, or crossing

crossing and crowding branches, and dead wood, permitting the general regular branches to advance according to their natural growth.

But the wall and espalier Pears having their branches constantly arranged horizontally, five or six inches asunder, retain the same branches many years as bearers, at full length, as far as the allotted space will admit, and as they will annually produce many useless shoots, they must be pruned every Summer and Winter.—Early every Summer go over the trees, and rub off all fore-right, and evidently too abundant and very rank shoots of the year, and train in only the main leading shoots and the best regular side ones entire, to chuse from in Winter pruning, if wanted.——In Winter pruning, performed from November till Spring,

Spring, that retaining the same bearing branches five or six inches distance, examine if any casually assume a decayed, or bad unfruitful growth, or are too crowded, or have advanced beyond their limited bounds, and prune them down to some lower bearer, if convenient; or occasionally retain regular young shoots of last Summer, advancing from below, at full length, to supply vacancies; and retain always the terminating or leading shoots to each horizontal entire, where room enough; cutting out all the superfluous, lateral young wood not now wanted, close to the main branches, leaving no stump, but carefully preserving all the proper natural fruit spurs; and cut out close, old and dead snags; [*See Apple tree*] then directly nail or tye in the general branches strait and close to the wall, &c. at regular distances.

The PLUM TREE.

THE Plum tree *(Prunus)*, furnishes a large collection of different varieties of its fruit, and is the original species of the family of *Prunus*, which comprises the Apricot, Bullace, Cherry, &c. as species of the same genus, agreeable to the botanic characters of the flowers and fruit; and belongs to the class and order *Icosandria Monogynia*, flowers containing twenty stamina and one style.

There is but one species of common Plum tree, which is called

Prunus Domestica; i. e. HOUSHOLD or COMMON CULTIVATED PLUM: *having oval, spear-shaped leaves, and the peduncles*

or footstalks of the flowers, for the most part *singly*, supporting white flowers of five petals, succeeded by the plums, which are round or oblongish, with a stone in the centre; and are of many different forms, sizes, colours, and qualities, in the numerous varieties; consisting of whites, blacks, yellows, reds, blues, and greens; ripening from July till the end of September, or beginning or middle of October, and of which there are varieties without end. But the most noted and approved sorts are known by the following names:

EARLY WHITE, or PRIMORDIAN PLUM. *Middle or end of July.*

EARLY BLACK DAMASK. *End of July.*

LITTLE BLACK DAMASK. *Beginning of August.*

ORLEANS PLUM, (red). *End of August and September.*

QUEEN CLAUDE, (green). *September.*

LITTLE QUEEN CLAUDE, (yellowish). *September.*

GREAT DAMASK VIOLET. *August.*

GREEN GAGE PLUM, (very fine). *August and September.*

BLUE GAGE. *September.*

WHITE PERDRIGON. *End of August and September.*

BLUE PERDRIGON. *End of August, &c.*

BLACK PERDRIGON. *End of August, or beginning of September.*

DRAP D'OR, or CLOTH OF GOLD, (bright yellow). *Beginning or middle of September.*

ROCHE COURBON, (red). *End of August, &c.*

WHITE BONUM MAGNUM, or EGG PLUM, (very large). *September.*

RED

Red Bonum Magnum, or Great Imperial Plum, (very large). *September.*

Fotheringham Plum, (large dark-red). *Beginning or middle of September.*

Brignole Plum, (yellowish). *September.*

Wentworth Plum, (yellowish). *Sept.*

St. Catharine Plum, (yellowish amber). *End of September.*

Royal Red Plum. *End of August and September.*

Cheston Plum, (blackish). *Middle of September.*

Mirabelle Plum, (greenish-yellow). *End of August, &c.*

Imperatrice, or Empress Plum, (dark red). *End of September, and beginning of October.*

Apricot Plum, (large yellow). *Beginning and middle of September.*

Inferior

Inferior Sorts.

PEAR PLUM, (whitish yellow). *September.*

LITTLE GREEN DAMASK. *Middle or end of September.*

MUSCLE PLUM, (dark red).

ST. JULIAN PLUM, (dark violet). *End of September, &c.*

DAMASCENE PLUM, (dark blue). *September and October.*

CHERRY PLUM, (small red). *Valued chiefly as a curiosity, but as it blossoms early is often cut off by the cold.*

The Plum tree grows fifteen or twenty feet high, branching with a moderate-spreading head, garnished with oval, spear-shaped leaves, and produces its flowers and fruit both on the young wood, from the eyes of the shoots, and on spurs

arising

arising on the sides and ends of the branches, of from two or three, to many years old; as in the Cherry, Pear, &c. the same branches continuing fruitful, and multiplying the bearing spurs many years, as they advance in length; and only require renewing with young wood occasionally, as any old bearer casually becomes unfruitful or decayed; being generally all permitted to extend in length, as shortening both destroys the first fruitful parts, promotes a great luxuriancy of useless wood from the lateral buds, and prevents their forming fruit-spurs.

This determines that we must not shorten the branches of plums in their general growth; besides too much use of

the

the knife on those trees occasions the branches to gum and decay.

The trees bear succefsfully in any order of training, either full or half standards, &c. and as wall and espalier trees; the standards planted twenty-five or thirty feet distance, permitting them to branch out freely, at full length, and form a full spreading head, pruning only any very irregular, rambling, or rampant growths; and the wall and espalier trees arranged eighteen or twenty feet asunder, with the branches trained down horizontally, five or six inches distance, without shortening, but extended as far as room will permit, especially all those designed as bearers, and they will emit fruit spurs all along their sides.

It

It is proper to have both a good share of common standards in the garden and orchard for the general supply; selecting the best sorts for walls and espaliers, in different exposures, which will furnish larger, earlier, and later fruit, and of an improved flavour.

They will all succeed in any common soil of a garden, orchard, &c. *See Planting*

Method of Propagation and first Training.

Plum trees, in all their varieties, are propagated, and continued always of the same sorts, by grafting or budding them upon any kind of plum stocks, raised either from suckers of the root, or from the stones of the plums, sowed in Autumn, two inches deep; and when a year old, planted out in rows a yard diſtance, to have from two or three, to four or five years growth; then graft or bud them with the proper ſorts, from four to ſix feet high, for half and full ſtandards, and within a foot of the ground for wall and eſpalier dwarfs, &c.

When the firſt ſhoots, from grafting, &c. are a year old, cut them down in March to a few eyes, to procure lower branches in order to give the head its firſt requiſite form, as in other fruit trees.

Afterwards let the ſtandards remain entire, and branch out each way: but the dwarfs for walls and eſpaliers may have ſome of the ſucceeding ſhoots ſhortened occaſionally, either in Spring, or young ones of the year pinched in Summer, to procure

procure a farther supply of collaterals to cover the wall and espalier regularly upward, with bearers, training them at full length.

Final Planting and General Culture.

Plum trees may be planted where they are to remain, when from one or two, to several years old, in Autumn, Winter, or Spring, with their heads mostly entire, except cutting out any very irregular growth; arranging the standards twenty-five or thirty feet distance, and the wall and espalier trees eighteen or twenty.

And as to general culture, observe nearly the following directions.

The standards should be permitted to branch out freely above, and form regular full heads, with the branches extending at their full length, only retrenching occasionally any very irregular and superabundant crowding growths, very long ramblers, and dead wood; with all suckers from the root and stem, and rampant shoots in the middle of the head: permitting all the general regular branches to extend as nature directs, without reducing their length; and they will soon be full of natural fruit spurs.

As to the wall and espalier plums, they must have their branches arranged horizontally to the wall and espalier, five or six inches distance, without shortening; and continue training, where necessary, a farther supply of new regular shoots of each future year, till they furnish the whole allotted space of walling, &c.

&c. completely with bearers, generally all extended entire; for if plum branches are shortened it will retard their bearing, and force out numerous rampant, useless, unfruitful shoots, and no fruit buds; but being arranged at their natural length, they shoot moderately, and in two or three years furnish abundant bearing spurs; the same branches should be continued many years as bearers, or as long as they remain fruitful, and only retain occasional new supplies of young wood, as any of the old casually assume a barren or decayed state.

But to preserve regularity and fruitfulness in the wall and espalier plums, they must be pruned every Summer and Winter.

Go over the trees every Summer, in

June

June and July, and difplace all fore-right young wood, evidently fuperfluous and very rank fhoots, with other ufelefs growths of the year, retaining only fome of the regular, moderate-growing fide fhoots, and main leaders, trained in at full length for occafional fupplies in Winter; continuing the whole clofe to the wall to admit the Sun and free air to the fruit.

Then in Winter pruning, continuing the fame bearers **five or** fix inches diftance, obferve **if** any difcover a bad growth, **or worn-out,** naked, unfruitful ftate, **which** may **now** be pruned or cut down to other more eligible **wood, or to any** contiguous young fhoots; retaining alfo regular fhoots of laft Summer, advancing below in vacancies,

cies, if any; and all such shoots that are not now wanted for the above purposes, should be cut out quite close, with all dead wood; preserving all the short natural bearing spurs on the sides of the branches, &c. but cut out too long, fore-right, projecting ones, old ragged snags and stumps; still retaining all the branches at full length, and let them be directly all trained in with due regularity to the wall and espalier.

The QUINCE TREE.

THE Quince tree (*Cydonia*), a species of the family of *Pyrus*, or Pear, but formerly constituted the genus *Cydonia*, is famous for its large, beautiful, golden-yellow fruit, of great fragrance, which, though too hard and austere to eat raw, is excellent for various culinary preparations; it therefore claims a place in the general collection of fruit trees, and is of the class *Icosandria*, and order *Pentagynia*, as the Pear.

There is but one species of the Quince, comprehending a few varieties, and is termed by the botanists,

Pyrus Cydonia,—CYDONIA, or QUINCE TREE; *having entire or unsawed oval leaves,*

leaves, hoary underneath, and whitish-red flowers placed singly, compofed of five petals, twenty ftamina, and five ftyles, fucceeded by large pyramidal and round golden-yellow fruit, having a hard four pulp, inclofing, generally, five hard kernels or feeds, and confifts of the following varieties, ripening in Autumn.

PEAR-SHAPED QUINCE.
APPLE-SHAPED QUINCE.
PORTUGAL QUINCE, with a tender pulp.
EATABLE QUINCE, having a tender pulp; fometimes eaten raw.

The Quince tree grows ten or twelve feet high, branching low and flenderly, within a moderate compafs, and produces its flowers and fruit fingly; fometimes on fmall lateral fhoots of the year, and upon small spurs from the sides of the older branches, which should generally be permitted to extend in length, without shortening them by pruning.

Quinces are valued principally only as a culinary fruit for stewing, baking, making marmalade, and to enrich the flavour of Apple-pies, tarts, &c. as being of a very heightened fragrance when fully ripe, but generally too hard and astringent to eat raw, as before observed.

A few trees should be arranged in the garden or orchard, chiefly as standards, and some also in espaliers for variety; planted eighteen or twenty feet distance; permitting the standards to branch out freely around, and they will bear abundantly; and train the espalier trees with their
branches

branches horizontally, at full length, fix inches afunder; managed as for Apples and Pears.

The trees are very hardy, and will thrive almoft any where; but they generally affect a moift fituation, and are therefore often planted along ponds, and ditch fides, in out grounds, &c.

Propagation and Training.

Thefe trees are expeditiously raifed from fuckers, cuttings, and layers, and occafionally by grafting, &c.

By SUCKERS.——Dig up the fuckers from the roots of any old Quince trees, in Autumn, &c. plant them in nurfery-rows, and train fome for ftandards, with four, five, or fix feet ftems; and others for dwarfs, headed down low to obtain lower branches.

By Cuttings.—Plant cuttings of the young shoots of the branches in Autumn or Spring, in a shady border; they will be well rooted by next Autumn; then plant them in open ground, to be trained as the Suckers.

By Layers.—Lay any convenient low-placed young branches in Autumn; they will root freely, fit to plant off next year, and managed as above.

By Grafting, &c.—Any particular desirable variety may be grafted or budded, either upon common Quince or Pear stocks, to continue and improve the sort, and have the trees sooner raised to a bearing state.

Planting

Planting and Culture.

Plant the standards twenty feet distance in the garden or orchard, or along the sides of ditches, or pools of water, &c. and let them branch into full heads, only retrenching occasionally very irregular or straggling growths, and permitting all the other general branches to remain entire.

And if any are designed for espaliers, set them fifteen or twenty feet asunder, the branches arranged horizontally, generally without shortening, and managed in general as advised for Apples and Pears.

The RASPBERRY SHRUB.

THE Raspberry *(Rubus Idæus)*, is an under-shrubby plant, four or five feet high, a species of the family of *Rubus*, or bramble, and produces agreeable eatable fruit of the baccaceous or berry kind, in estimation both as a desert fruit to eat raw, and for making tarts, sauces, raspberry jam, and other culinary preparations, and therefore highly demands culture in every garden: it belongs to the class and order *Icosandria polygynia*, flowers having twenty or more stamina, and many styles.

There is but one species of the common fruit-bearing Raspberry, furnishing several varieties, and according to the botanists

botanists bears the following name and description.

Rubus Idæus, COMMON RASPBERRY PLANT; *having prickly stalks, pinnated or winged, five and three-lobed, rough leaves, on channelled footstalks,* and clusters of white and purple flowers, of five petals, with numerous stamina and styles, succeeded by roundish red and white soft berries, composed of many acini, and consists of the following varieties, ripening in July and August, viz.

COMMON RED RASPBERRY. *July and August.*

COMMON WHITE RASPBERRY. *July, &c.*

SMOOTH-STALKED RASPBERRY. *July, &c.*

TWICE-BEARING RED RASPBERRY. *July, and again in September and October.*

TWICE-BEARING WHITE RASPBERRY. *July and September, &c.*

Of the above varieties we recommend the firſt two or three ſorts for general culture, for the main ſupply, as being generally the moſt plentiful bearers; but the twice-bearing ſorts ſhould alſo be admitted in every collection, for they likewiſe bear tolerable crops of good fruit; and what is remarkable, they produce two crops every Summer; the firſt in July, and the ſecond in September, &c. often in tolerable good perfection.

The Raſpberry plant grows four or five feet high, ariſing with many ſlender, erect, prickly, and ſmooth ſtems, immediately from the root, annually; of an under-

under-shrubby herbaceous nature, as although they become somewhat ligneous or woody; they are not durable, being only annual, or, at most, biennial; rising from the root one year, and the next emitting many small lateral or side shoots, bearing the fruit the same Summer, then totally die to the root in Winter following; succeeded, however, always by a plentiful succession of young stems from the same root or stool in Summer, for next year's bearing; every winter the old decayed stems, which bore last Summer, are retrenched to the bottom, to give place to the young successional supply, which, at the same time, are thinned to from three or four, to five or six of the strongest, on each main root or stock, and their weak tops generally pruned down

down a foot, or more, if very long, to render them more robust, erect, and regular, as well as to promote a more plentiful supply of collateral shoots in Summer, for the immediate bearers.

For the flowers and fruit are always produced on the shoots of the year, emitted from the sides of the main stems, as aforesaid, and at the axillas or angles, formed by the stem and collateral shoots; gnerally terminating the shoots in clusters, flowering in June; and the fruit ripens in July and August.

They are very hardy plants, that prosper any where in a garden, or any open ground; planted generally in rows four feet and an half distance, and a yard in the row; and sometimes also disposed singly

gly in borders, verging walks, and in shrubbery compartments, &c. for variety; generally permitted to advance with several stems from the same root, in a bushy manner.—*See the General Culture.*

Sometimes, for variety, a few are planted espalier ways, and the shoots arranged at full length to stakes.

A plantation of Raspberries continue bearing plenteously for several years, renewing their stems annually, as before remarked; though, I would observe, that a plantation of more than five or six years standing, generally produces smaller fruit, and of an inferior quality to that on younger plants; it is therefore proper to plant them afresh in another plat of ground every four, five, or six years, from young suckers or stems, as below.

Method of Propagation, Planting, and Culture.

The propagation of Raspberries is effected with facility and expedition by the suckers or stems, arising abundantly from the root, annually in Summer, forming proper plants for planting out in Autumn, Winter, or Spring following, and will bear fruit the ensuing Summer.

The method is this:

Having, in Autumn or Winter, &c. fixed on an open spot of good ground, well dunged, and prepared by proper digging, &c. then proceed to procure the plants from a plantation of good well-bearing Raspberries, chusing a quantity of the strongest outward young suckers, dug up with as many fibres as possible; pruning their long straggling roots, and

and any naked woody knobs, part of the old stock, that may adhere; preserving any young advancing buds at foot of the stem, for future shoots or succession bearers: and pruning the top of each plant to about a yard long, ready for planting.

Then proceed to plant them by line and spade, in rows, South and North, if convenient, for the greater advantage of the sun; placing them a yard distant in each row, and the rows a yard and an half, or five feet asunder, each plant in an upright position: and if dry ground, and late Spring planting, give each a good watering, at bottom, to settle the earth, and forward their taking fresh root.

Thus they will soon readily take root, pro-

produce shoots at top, and bear fruit the ensuing Summer, as well as send up each several succession stems from the bottom, and form a full plantation for bearing plentifully the succeeding year.

Then, as to general culture, observe as follows:

In Summer, keep them clean from weeds, by occasionally hoeing the ground in dry weather, pulling up all widely-straggling suckers of the plants, arising between the rows, &c. and reserving a sufficiency of the strong shoots about each main stock or stool for succession bearers, to produce the next year's fruit.

Every Winter, any time from November till March, they must have a general dressing, which consists in retrenching all the old stems, or last Summer's bearers, as useless, they not surviving the Winter to bear again: selecting, at the same time, a supply of the strongest young stems on each root, to furnish next year's fruit, and thin out the superabundancy: proceed therefore to cut or break down all the old stems close to the ground; and select from three or four, to five or six of the strongest, best-placed, young shoots on each stock; cut out also all the others close to the bottom, together with all stragglers between the main plants, and let each of the reserved shoots have its weak or bending top pruned, to render them more robust and strong, to support their upright position in Summer, as well as to encourage a stronger production of lateral twigs as the immediate bearers, as before explained.

Though

Though sometimes we shorten the stems but moderately, or only just down to the bend or weak part at top, and sometimes not at all; and if they are long, and stand stragglingly wide, plait them together by threes, &c. or in an arched manner at top, and they will thus support one another upright.

But, for variety, may train some rows to stakes arranged espalier ways, as before observed, laying the shoots horizontally, nearly at full length, six or eight inches asunder, and they will thus often produce larger fruit, earlier ripe, and with an improved flavour.

As soon as the Raspberries are Win-dressed, clear off all the cuttings and rubbish, bith, and let the ground be digged one spade deep; and as you proceed, dig up all ſtraggling ſuckers and roots not belonging to the main ſtools.

A little rotten dung applied once in two years between the rows, at Winter-dreſſing, and digged in, will prove beneficial in ſtrengthening the plants, as well as improve and prolong their fruitful ſtate, and promote the ſize and quality of the fruit.

The SERVICE BERRY TREE;

Or, WILD SERVICE.

THE Service-berry tree, or Wild Service *(Cratægus)*, grows wild in England, attaining a lofty stature, and is often introduced in gardens and orchards as a fruit tree, for the sake of its berries, which grow in large brown bunches; and, when fully ripe and soft, in Autumn, have an agreeable tartish flavour, and makes a variety among the late fruits; and is of the class and order *Icosandria Digynia*, twenty or more stamina and two styles, to each flower.

There

There is but one species that claims attention as a fruit tree, and which the botanists entitle,

Cratægus Torminalis,—EATABLE, or MAPLE-LEAVED WILD SERVICE—*Having heart-shaped, seven-angled leaves, with the lower segments spreading asunder;* and large bunches of white flowers, of five roundish petals, containing many stamina, and two styles, succeeded by clusters of reddish-brown berries, having two hard seeds, ripening in Autumn, which, after being gathered and lain till they become soft, are palatable to eat.

This tree grows forty or fifty feet high, with a large, branchy, spreading head, ornamented with large heart-formed, sept-angular leaves, hoary underneath;

neath; and produces its bunches of flowers and fruit on long foot-stalks, towards the upper part and ends of the younger branches.

It grows wild in woods, in England, &c. but merits cultivation in gardens and orchards, as a fruit tree, trained chiefly as full or half standards, arranged in assemblage, or dropped singly in shrubberies, parks, or lawns, &c.

For variety, some may also be trained as dwarfs, arranged in espaliers, and managed as directed for Apples, Pears, &c.

Method of Propagation, &c.

It may be propagated by the seed or brries, by layers, and grafting and budding.

By

By Seed.—Sow the berries in Autumn or Winter, or early in Spring, in a bad of light earth, in drills two inches deep; and as they will probably not all rise till the second Spring, keep the bed clean from weeds all Summer, &c. and when the plants are a year old, plant them out in the nursery, and train them with single clean stems, from four to six or seven feet high, for half and full standards; then let them branch out above, and form heads.

But if you design any for dwarfs, should head them when young, near the ground, to gain lower branches, managing them as other dwarf fruit trees.

By Layers.—Where any of the trees furnish low branches, or have been headed ed down as stools, to produce shoots for layers, near the ground, lay them in Autumn in the usual manner, and by next Autumn they will be rooted, then plant them off, and train them as the seedlings, directed above.

By Grafting, &c.——By this method the trees will sooner arrive to a fruitful state, and may be performed upon seedling stocks of their own kind, or hawthorn, or any species of the *Cratægus*, or Wild Service stocks, in the usual method, both for standards and dwarfs, as directed for Apples, &c.

Final Planting, &c.

When the trees are from five or six, to seven or eight feet high, as standards, they may be planted where they are to remain, in the order before observed; and

and permitted generally to assume their own mode of growth, except reforming any casual irregularities, as in other standard fruit trees.

And for espaliers, they being previously trained with low stems, branching near the ground, may plant them eighteen or twenty feet distance, arranging the branches horizontally, mostly at full length, six inches asunder, and managed as for Medlars, Pears, &c.

The SORB TREE,

Or CULTIVATED SERVICE.

THE Sorb, or cultivated Service, (*Sorbus*) is of a separate family or genus from the Wild Service, by having three styles and three seeds, the other but two; and the fruit is considerably larger, being the size of little apples, but also of the baccaceous or berry kind, and somewhat of the nature of Medlars, in not being good to eat till it assumes a state of decay, in Autumn, and belongs to the class and order *Icosandria trigynia*, i. e. twenty or more stamina and three styles in each flower.

There

There is but one species meriting culture as a fruit tree, viz.

Sorbus Domestica, HOUSHOLD, or CULTIVATED SERVICE TREE.—*Having pinnated or winged leaves*, of many pair of lobes, and an odd one; *hairy underneath*, and large bunches of whitish flowers, of five roundish petals, succeeded by Pear and Apple-shaped reddish fruit, in clusters, containing three or four seeds, and consists of the following varieties, ripening in Autumn.

PEAR-SHAPED SERVICE:
APPLE-SHAPED SERVICE.

The trees grow thirty or forty feet high, adorned with winged leaves, and produces the flowers and fruit at the ends and sides of the younger branches, and on lateral shoots or spurs, generally in bunches, flowering in May or June, the fruit ripens in September, which being then gathered, and deposited in the fruitery a little time, to become mellow, they will eat with an agreeable relish, effecting a variety among the Autumn fruits.

A few trees of this sort, trained as full or half standards, may be ranged in the garden or orchard, or dropped singly in different compartments of the shrubbery, &c. and may likewise, for variety, and to improve the fruit, train some in espaliers, in concert with Medlars and Quinces, &c.

Method of Propagation and Training.

It may be propagated by seed; but to continue the desirable varieties distinct and

and permanent, they should be propagated by grafting and budding.

By SEED.—Sow the seeds of the fruit in Autumn, either in a warm border, two inches deep, or in pots, to move under shelter of a frame in frosty weather; but if plunged in a hot bed in Spring, it will forward the germination of the seeds; but those in the full ground will also grow, though not so forward as the others; plant them out in Autumn or Spring following, in nursery-rows; and train the principal part as half and full standards, with from four to six or seven feet stems, branching out at top, and forming spreading heads: some may be trained for espaliers, being headed near the ground, at one or two years old, to gain lower branches, as directed for Almonds, Apples, &c.

By Grafting and Budding.—The approved varieties are with certainty continued by this method; worked either upon their own seedling stocks, raised as above, or upon Pear stocks, and trained for the purposes intended, as directed for Apples and Pears.

Planting and Culture.

They may be planted as standards, when about five, six, or seven feet high, having formed proper heads, which permit generally to take nearly their own growth, like other common standards.

Or if you raise any dwarfs for espaliers, for variety, plant them eighteen or twenty ty feet asunder, with their branches ranged six inches distance, retaining young laterals, occasionally coming up anew, to supply the places of old, worn-out, or too long advanced bearers, giving a Summer and Winter dressing, as for Apples and Pears, &c.

The VINE;

Or, GRAPE VINE.

THE Vine, (*Vitis*) claims precedence of most other fruit trees, for the great and rich variety of its most excellent and valuable fruit, the Grape, universally celebrated for its deliciously rich juice, so eminent for making wine; and is also a delicately fine eating fruit, of the richest flavour; it is of the berry kind, growing in large long clusters, and comprehends numerous varieties, all the progeny of one mother species; which belongs to the class and order *Pentandria Monogynia*, i.e. five stamina and one style in each flower.

There are several species of Vine; but the principal noted sort, valued for its fruit, the botanists distinguish by the following name, &c.

Vitis Vinifera; the VINE, or GRAPE TREE—Described, *Vine, with large angulated, lobated, sinuated, naked leaves*; having claspers arising opposite the base of the footstalks, and clusters of very small greenish flowers of five petals, five stamina, and one style, succeeded by large long bunches of roundish or oval berries, of different colours in the varieties, ripening from July till October; some sorts black, others white, red, &c. as below.

The principal varieties are generally known by the following names:

BLACK

Black July Grape. *Beginning or middle of August.*

Black Sweet-Water Grape. *Middle or end of August.*

White Sweet-Water. *Middle or end of August.*

Black Cluster Grape, having hoary, whitish leaves, and short compact clusters of grapes. *September.*

Early White Muscadine. *Early in September.*

White Muscadine Royal, or Chasselas Blanc. *September and October.*

White le Cour Grape, or Musk Chasselas. *September.*

Red Chasselas. *September and October.*

Black Corinth, or Currant Grape, (small). *August and September.*

Black Burgundy Grape. *September and October.*

Red Hamburgh Grape, (large). *October.*

Black Hamburgh Grape, (large). *October.*

Black Frontinac Grape. *End of September and October..*

The following, being mostly fine, large, late ripening grapes, and some of them exceeding large bunches, unless the Autumn season proves very warm and dry, do not ripen freely in England, and are therefore often planted against hot walls, hot-houses, and forcing-frames, &c. to obtain them in the utmost perfection early in Autumn.

Red Frontinac, (large and rich). *September and October.*

GRISLY FRONTINAC, (large and fine). *September and October.*

WHITE FRONTINAC, (large and rich). *September and October.*

WHITE MUSCAT OF ALEXANDRIA. *End of September and October.*

RED ALEXANDRIAN MUSCAT. *End of September and October.*

ST. PETER'S GRAPE, (black, and very large berries and bunches). *October.*

TOKAY GRAPE, (white and very rich). *October.*

WHITE SYRIAN GRAPE, (exceeding large clusters. *October.*

RED RAISIN GRAPE. *End of October.*

WHITE RAISIN GRAPE. *End of October.*

CLARET GRAPE. *October.*

Of the above varieties, most of the first ten

ten or twelve sorts being trained against good sunny walls and buildings, ripen freely in all warm dry Autumns, and sometimes in espaliers, and in vineyards; but the other sorts do not always attain perfect maturity without artificial heat, as aforesaid.

The Vine is a flexuose and climbing plant, unable to elevate itself erect without support.

It grows with long, flexible, woody stems, many feet high, by support, making numerous shoots, three or four yards long, or more, in one Summer, consisting of many long joints, garnished with large, angulated leaves, attended by tendrils twining round any thing they encounter; and produces the flowers and fruit on the young shoots of the same year, arising in clusters from the buds of the shoots; but from such shoots only, that arise immediately from the former year's wood, for Vines rarely produce immediate-bearing shoots from the old branches, or from any but the shoots produced the year before, so that a general supply of every year's shoots must be retained in every part, both in Summer, as the present bearers, and in Winter dressing to furnish the succession bearers next year, &c. trained mostly at full length all Summer, and shortened to a few eyes always in Winter pruning:

For in Vines the same individual shoots never bear but once; but the bearers of each year producing a succession of of numerous shoots, each succeeding Spring and Summer, on these only the same year's fruit is always produced; and in this manner the succession of bearing wood is continued.

All the varieties of Vine require cultivation mostly as wall trees, against warm southerly walls, and occasionally in well-exposed espaliers; otherwise the Grapes will not attain due perfection; so should generally allot the principal supply against good South walls, pales, or buildings, open to the sun, and occasionally on East and West walls, in default of sufficient scope of South aspects.

For this purpose they are commonly trained with low stems, half a foot or a foot high, branching out low, in order to occupy the whole wall quite from the bottom; training the branches either horizontally or erectly, as the scope of room, and height of the wall admits; ranging the main branches a foot or more asunder, and retaining lateral young shoots of each year between, advancing behind one another from the bottom to the extremities, as the principal bearers; and according as the old branches advance too long for the wall, they must be cut down to lower shoots in Winter, and the supply of shoots then retained, should be shortened, as directed in their *General Culture*.

Some also of the forward Vines may likewise be planted in espaliers, as also in the vineyard way, arranged to stakes; and in favourable seasons they will ripen tolerably good crops of Grapes.——— *See their Vineyard Culture*.

But sometimes, in unfavourable wet Summers and Autumns, Grapes ripen but indifferently, even against the best walls, more especially the larger late Grapes; and sometimes not at all in bad seasons, for which reason, some of the choicer forward and late sorts are often planted against hot walls, or forcing frames, furnished with ranges of flues for fires, and defended with glass frames in front; or sometimes planted against the front or end walls of common hot-houses, and the branches introduced, by conducting them through small holes, and trained up against the inside glasses or walls, and thus obtain early Grapes two or three months before the natural season in the open ground; and have the late kinds ripened in the utmost perfection.

This is well worth practising by every one accommodated with the above conveniencies, or if only with a common Pine Apple stove.

However, it is also proper to plant some, both of early, middle, and late kinds, against common walls or pales, &c. in the open ground, to take their chance; and if their branches are kept always thin, and the shoots trained in close all Summer to admit the full sun, there will be no fear of success in all favourable seasons.

Vines are very hardy in respect to growth, and succeed in any good garden earth; but in dry, light, warm soils they are more successful in ripening the Grapes earlier, in due perfection, and richness of flavour; so that in soils naturally strong,

strong, or of a clayey moist temperature, the border may be improved with light dry materials, such as any light sandy or stoney earth, sea sand, road stuff, lime rubbish, coal ashes, &c. worked both in a stratum at bottom, and blended with the common earth of the border, where the Vines are to be planted.

Method of Propagation and Training.

Vines are propagated principally by layers and cuttings of the young wood of one year old, and raised to a state of bearing in two or three years; though layer-raised plants often bear the ensuing Summer.

By LAYERS:—In Autumn, Winter, or Spring, chuse some strong lower shoots,

shoots, or young branches furnished with such, and opening an aperture in the ground, longitudinally, five or six inches deep, deposite the body of the shoot or branch therein, and pegging it firmly down, cover it with the earth, and shorten the top shoots to three or four eyes; they will thus readily take root below, and shoot out at top in Summer; and in Autumn following they may be planted out either to remain, or in the nursery till wanted.

Then, in Winter, they having made some strong shoots the preceeding Summer, let them be pruned to three, four, or five eyes, and trained to the wall, &c. and as they will shoot strongly in the Summer following, from each remaining eye, train the new shoots mostly entire,

till

till next Winter, then prune and train them as before, and they will bear the second year. *See the General Culture.*

By Cuttings.—The young shoots of the former Summer are the only proper parts for cuttings, planted in Autumn, or early in Spring; chusing principally the lower and middle parts of the shoots as the strongest best-ripened wood, cut into lengths of about three joints: and then planted either where they are finally to remain, as observed of the layers, in a shady border, or in nursery lines a yard asunder, and a foot in the row; planting them almost down to the top, leaving only one eye above ground, and that almost close to the surface.

Keep them clean from weeds, and give

give waterings in Summer, when they will readily emit roots below, and shoots at top, which retain at full length till Winter; then shorten them to two or three eyes, and manage them as directed for the layer-raised plants, and as in the *General Culture.*

Final Planting in the Garden, &c.

Vine plants, from one or two, to several years growth, may be succesfully planted, though young plants are preferable to old, any time from November till March, in open weather.

Let them be digged up with good roots; trim off only straggling or broken parts thereof, and prune the shoots of the head to about three eyes, if quite young plants,

plants, and to three, four, or five, &c. in older Vines; then plant them along the wall, &c. in the common method of planting, ten or fifteen feet distance if for a full plantation; or if straitened for room may plant some in the vacant intervals, between Peaches and Nectarines &c. giving them a moderate watering to settle the earth, and promote their rooting; then directly fasten their shoots to the wall or espalier, ten or twelve inches asunder.

The General Culture, &c.

In the general culture of Vines, observe they may be trained either horizontally or upright, as the space of wall admits; and that in respect to pruning and training, as they shoot numerously every Summer, the general mother branches
should

should be ranged ten or twelve inches distance, at least, in order to have sufficient room to train the requisite supply of the immediate-bearing shoots of each Summer, and other regular-placed ones, for next year's bearers.—And as they every year produce a more numerous supply of shoots than can be trained or converted to use, consistent with regularity, they accordingly require a regulation of pruning and training every Summer and Winter.

In Summer should go over the Vines frequently to regulate the growths of the year; commencing the first regulation early in May, or as soon as the fruit-shoots discover the fruit buds, and before the general shoots run into confusion, which would occasion much anxiety

and

and perplexity to regulate, as well as prove detrimental to the growth of the Grapes, which should have every possible advantage of the sun, to accelerate their perfection.

Therefore, in this season, (Summer) begin early to rub off all the very weak, straggling, evidently unfruitful, and other useless shoots of the year, and particularly such as rise directly from the old wood, unless required to supply vacancies, retaining all the good fruitful shoots, discovering the bunches of flowers in infancy, also a sufficient supply of other well placed strong shoots, to have plenty to chuse from in Winter pruning, for next year's mother-bearers; at the same time displacing all others that are obviously superfluous

superfluous or unnecessary, and all small laterals arising on the sides of the reserved supply, which, when long enough, train in at full length; or, in July, may top the present fruit-shoots, to throw a greater supply of nutriment to the fruit; but shoots not furnished with Grapes may generally be extended as far as you can, for if shortened too early in their growth, they will force out numerous useless lateral shoots, crowding the Vines, and prove hurtful to the buds from which we are to expect the next year's fruit shoots; so should generally run them as far as they have room, till Winter pruning.

After performing the general Summer pruning, and dressing the Vines, we should review them every week or fortnight, to adjust casual irregularities, displace

place all after-shoots, and to train the regular supply along close to the wall, &c. as they advance in length, in order both to preserve the requisite uniformity, and to admit more freely the necessary benefit of the sun and air, to promote and improve the growth of the fruit, which, in this country, requires every possible advantage, to forward and ripen it in due perfection.

The Winter pruning may be performed any time, from the fall of the leaf till Spring, observing we are now to regulate both the young supply of future-bearing wood retained in Summer, as well as the old bearers, and long naked branches, some of which should always be pruned away annually, in Winter, down to lower branches, to make room

room to train the young succession bearing shoots.

Therefore, selecting a general supply of the most eligibly-placed strong shoots of last Summer, arising principally on the year-old wood, to retain for next year's mother bearers, advancing at proper distances regularly from the bottom upward, between the older branches in successive order, one behind another; retaining generally one on each former year's branch, or sometimes two in wide spaces, if necessary, and prune out all the superabundant ones, with all weak and irregular shoots quite close, together with part of the former bearers, pruned down to their respective successional shoots, now proper to retain; as likewise long naked old branches,

ches, not furnished with young wood, or that are advanced to the top of the wall, cutting them down to the best lateral branches, or shoots they support, thereby making room to train the general supply of young bearers ten or twelve inches distance, with one forming a leader to each main branch, and some always advancing from below: as you proceed, let all the reserved supply of shoots be shortened, from three or four, to six eyes or joints, or more, according to their strength and situation on the Vine; or if you want to run them along any vacant or high space, may be occasionally left longer in proportion; performing the shortening just above an eye, with a sloping cut upwards.

As soon as pruned, nail the whole close

to the wall, &c. either inclining horizontally, or more afcending, as the fcope of walling admits, arranging them ten or twelve inches diſtance, at leaſt, to allow full fcope to train the enſuing Summer's ſhoots between.

Where any part of the bottom of the wall is naked of bearing wood, or that you would extend the Vines farther, may lay down any convenient branches in the earth, and they will root and fupply the places required.

Of Eſpalier Vines.

When deſigned to have Vines in eſpaliers, plant them in funny expofures, and manage them as directed for the wall Vines; one mode of culture will fuit in both kinds of training, or as below for the Vineyard.

Of the Vineyard Vines.

Vineyards are plantations of Vines, cultivated in the open ground, without the assistance of walls, or any other close fences, but arranged in several parallel rows, cross-ways the ground, eight or ten feet distance, one row from the other, and the branches of the Vines trained along to stakes, ranging the way of the rows, espalier ways; and are designed principally to produce large quantities of Grapes for making Wine.

This is the common method abroad, in the wine countries, of cultivating their very extensive plantation of Vines for affording the general vintage of Grapes for the purpose of making wine as aforesaid.

It

It has also been attempted in England, in moderate plantations, but not with such general success as abroad, in warm climates, our autumn seasons not proving always favourable enough to ripen the Grapes in any due perfection; it, however, is worth the trial, as Vines will often bear abundantly in this order of training; and we have also had them ripen in tolerable good perfection, especially when growing in a warm dry soil, and southern exposure, open to the full sun.

Abroad they often plant their vineyards, on hilly, or on any elevated situation where the soil is dry and warm, and to obtain this they often plant on stoney, rocky, chalky, or gravelly places, that the warmth of the soil may contribute

tribute to forward the ripening and improving the flavour of the Grapes; and such situations and soils have also been employed in England for the same purpose, with tolerable success.

However, they may be planted to form Vineyards in any common dry soil, not too clayey or wet, and that lies well to the full sun, from rising to setting, or on the side of a moderate acclivity, facing the South; preparing the ground by proper trenching, or deep ploughing, &c.

The plants for this purpose may either be layers, raised as we before directed, or cuttings, either planted at once to remain, as before explained in the garden culture, or planted in the nursery, and trained tr ined a year or two; then transplanted into the vineyard.

Observing, however, in planting them in the vineyard, to set them six or eight feet distant in the row, with an interval of ten feet between the ranges.

Observe, likewise, if at the time of planting them they are furnished with one, two, or more shoots; let each be pruned to about three buds or joints, in length, and fasten them to short stakes placed in the ground for that purpose, as below.

Then, as to their general future culture, the branches must be constantly trained to stakes, arranged along each row of Vines, about three feet high, at first; but as the Vines advance in age, strength,

strength, and number of branches, must have higher and stouter stakes; training the shoots to the stakes with osiertwigs, &c.

In Summer, when they shoot forth, clear off all small twigs arising on the main shoots of the year, and other weak useless growths, training the main young shoots along to the stakes, at full length all Summer: in Winter shorten them to three eyes, and when they shoot again, in Summer following, manage them as before, continuing to encrease the main branches to six or eight on each plant.

But when the Vines are three years old, or more, having six or eight principal branches, and bear tolerably, may top the young bearing shoots of the year, in June, to two or three joints a-

bove

bove the fruit, to forward and ſtrengthen its growth, clearing off all lateral twigs, and uſeleſs ſuperfluous ſhoots, and training the others cloſe to the ſtakes; in Winter pruning, ſelect ſix, eight, or ten of the ſtrongeſt, beſt-placed young ſhoots of laſt Summer on each plant, for next year's bearing, pruning out the ſuperabundancy, as in the garden Vines, and ſhortening the reſerved ſhoots; the ſtrongeſt ones cut to four or five joints, and the weaker prune to three eyes each; then faſten them along regularly to the ſtakes, an equal number to the right and left, ten or twelve inches aſunder; thoſe next the ſtem may be trained nearly upright, but the lower ſhoots ſhould be extended more horizontally.

Every Spring, after Winter pruning and

and dressing the Vines, the ground between the rows should be neatly digged, and about once in three or four years add some manure, either of good rotten dung, or a compost of dung and fresh earth, lime, &c. applying it at the above season, digged in one spade deep.

All Summer keep the ground very clean from weeds, by frequently hoeing it in dry weather, to preserve a clean dry surface, in order both to admit the sun freely to the ground, and to reflect its heat more powerfully on the fruit, which will contribute considerably towards improving its growth, and enriching its flavour.

The WALNUT TREE.

THE Walnut (*Juglans*), is one of the largest fruit trees of the nut-bearing kind, and is worthy of culture as common standards, in orchards, parks, and any out grounds for the sake of its fruit, which is valuable, both to use whole, while young and green, as an excellent pickle, and when ripe, to eat the kernels raw; consisting of several varieties, all seminals of one parent, which belongs to the class and order *Monoecia Polyandria*, i. e. male and female flowers apart, the males having many stamina.

The species of Walnut commonly cultivated as a fruit tree, is, according to the botanists,

Juglans

Juglans Regia, COMMON WALNUT TREE; described *Juglans, with winged leaves, of five or seven large oval, nearly-equal, smooth lobes*, and with small six-parted male flowers, in oblong, scaley catkins, and females in close-fitting clusters, succeeded by large, oval and roundish green fruit, each including one large, oval, furrowed nut, containing a four-parted, eatable kernel, ripening in September and October, and comprehends the following varieties.

EARLY OVAL WALNUT.
COMMON OVAL WALNUT.
ROUND WALNUT.
LARGE WALNUT.
LARGEST FRENCH WALNUT.
LARGE DOUBLE WALNUT.
LATE-RIPE WALNUT.

TENDER-

TENDER-SHELLED WALNUT.
HARD-SHELLED WALNUT.

The Walnut tree grows forty or fifty feet high, branching out widely around, garnished with large pinnated leaves, and produces the flowers and fruit near the ends of the former and same year's shoots, towards the extreme parts of the branches, growing generally in clusters; the flowers appear in April and May, succeeded by the fruit in June and July, gradually encreasing in growth till September, when the Walnuts begin to ripen; and when arrived to full perfection, the green outer cover divides and discharges the nut, containing the eatable kernel.

As we above hinted, the fruit of the Walnut tree is useful in two different stages of growth, viz.

When

When green, young, and tender, in July and August, about half or three parts grown, is excellent for pickling, using them whole, the outer cover, shell and kernel together, before the shell becomes hard, when they make a very fine high-relished pickle for use the year round.

And when fully ripe the latter end of September, and in October, the kernels being of an agreeable bitterish flavour, are exceedingly palatable to eat, and continue in perfection six weeks, or two months, or may be kept double that time in a dry room, closely covered with straw or dry sand.

So that the Walnut, considered as a fruit tree, highly deserves a place in the collection,

collection, trained as common standards, in orchards, parks, avenues, and the borders of fields, or on any out parts, either in continued ranges, forty or fifty feet asunder, or more, where large quantities of the fruit are required for any public demand, or in detached standards, singly, here and there, or arranged on the boundaries of orchards, &c. in assemblage with Chesnuts, where they will also defend the interior more capital fruits from the insults of boisterous winds, in all of which they should generally be employed as full standards, with six or seven feet stems, and suffered to branch out above into spreading heads, without shortening the branches, as they bear always mostly towards the extremities.

Though

Though for private use, a few trees are sufficient, yet, where there is large scope of ground, I should advise having plenty of them disposed in different situations, as they will not only be profitable in their annual crops of fruit, which is always ready sale in the markets, but will effect variety and ornament in their growth; and when arrived to timber prove valuable for many purposes in the cabinet and joinery branches.

It is a very hardy tree, and will succeed in either low or high situations, and almost any soil, but is the most prosperous in loamy ground.

Propagation and Training.

This tree is raised most commonly from the nuts; though as the seedlings are apt to degenerate or run to different sorts, we cannot be certain of continuing the varieties distinct.

However, should be careful to provide a quantity of the best, large, thin-shelled nuts, with well-flavoured kernels, when thoroughly ripe, to preserve in dry sand till February, then plant them in any lightish ground, in drills three inches deep, and a foot asunder, and they will come up in a month or two, and grow half a foot high, or more, the same year; and in the first or second Autumn or Spring after, be fit to plant out; previously, when taken up, shorten the downward top root, and plant them in nursery lines a yard distance: here train them for full standards, with single clean stems, six feet high or more, then permit them to branch out at that height, and form full heads, after which they may be planted out finally to remain.

They rarely begin to bear till seven or eight year's old, but not considerably till they attain a large growth.

Final Planting, &c.

The planting of Walnut trees may be performed when they are from six to ten or twelve feet high, though if not more than six or eight feet they may prove more successful: the proper season is either at the fall of the leaf, or the following months, till March, taking them up with good roots, of which trim off only broken parts, and leave the top entire, then plant them where required, thirty or forty feet distance, and let each be staked, to prevent their being disturbed by tempestuous winds.

As to culture, hardly any is required; let them generally branch out all around, according to their natural growth, except occasionally to lop any very irregular bough, low straggler, or very long rambling branch.

When designed to gather Walnuts to pickle, July and August is the time, before the shell is hard; chusing such that are as free from specks as possible, and gathering them carefully in a dry day, by hand, without bruising.

As to the ripe fruit, they are ready, some the latter end of September, others not till October; when those on small trees may be easily gathered by hand, as wanted, but on trees with high, and widely-extended heads, they are commonly beat down with long poles; and as the outer husk or cover generally adheres close, they should be gathered up in heaps, to ferment and sweat a few days, when the green covers will separate from the nuts, which being then cleaned from the rubbish, lay them up in a dry room, or in boxes or tubs of sand, for use.

GENE-

GENERAL OBSERVATIONS.

Of Grafting and Budding.

THESE methods of Propagation are effected by the insertion of a young shoot or bud of any desirable variety, into the stock or stem of another of the same genus or family, or at least of one nearly related to it; and these uniting, shoot forth into branches; forming the new tree of the intended sort, producing fruit in all respects like to that of the parent one; thus our choice varieties of fruits are encreased and continued.

The first thing to be observed respects the proper stocks. Apples succeed only upon Apple-stocks; Pears upon Pear-stocks, or occasionally on Quinces, as being of the same *genus*.; so of Plums and

and Cherries, and most other fruit trees except in Peaches, Nectarines, and Almonds, which, though of the genus *Amygdalus*, succeed upon the *Prunus* or Plum-stock, which is of the same class and order, and the favourite stock, in being more hardy and durable than those of their own kind: It must likewise be observed, that though the several species of any genus will succeed upon one another, yet there are particular sorts in the same family more peculiarly adapted to their own species; for instance, the Apple and Pear being different species of the same genus, will grow upon each other, but not prosperously: Plums and Cherries may be produced one on the other, but not near so successfully as each on its own stock, and so of many others.

The

The methods of raising the different stocks are by seed, suckers, layers, and cuttings, but principally from the seed, kernels, or stones of the respective fruits, sowed in Autumn or Spring, in beds of light earth, one or two inches deep. At a year old they are to be planted out in nursery-rows a yard asunder, where, in a year or two after, many will be fit to graft or bud to form for dwarfs wall and espalier trees, and in three, four, or five years for standards.

Grafting and Budding are also occasionally performed on trees which already bear fruit, with design either to change the sorts, or have two or more kinds of fruit on the same tree.

The months for Grafting are February and

and March, beginning with the earlieſt ſorts, and ending with the late-ſhooting trees, ſuch as Apples, &c. the ſhoots for cions or grafts to be collected in February, before their buds advance too much. Theſe ſhoots ſhould be the production of the former Summer, moderately ſtrong, robuſt, and clean, tie them in bundles, and place their lower ends in earth for uſe as they are wanted.

The denominations of grafting commonly practiſed, are Whip-Grafting, Cleft-Grafting, and ſometimes Crown-Grafting, but chiefly the firſt of theſe, being the moſt expeditious and ſucceſsful.

Whip-Grafting is adapted to ſmall ſtocks, about half an inch thick, if the
ſtock

stock and graft are nearly of a size, the grafting will succeed the better. Cut off the head of the stock, at the height intended to form the stem, with a slope near two inches long, make a thin slit downward, from the top of the slope, about half an inch, then shorten the cion at top to five or six eyes, and cut the lower end sloping; make also a small slit or tongue near the top of the sloped part, upwards, so as to fit, being careful, however, in tongueing the graft, not to go too deep towards the back of it; apply the tongued part of the graft in the slit of the stock even and close, bind them with a ligature of bass several times round, and secure the whole with a coat of clay an inch thick, and an inch above and below; finishing in a rounding form, so perfectly close, that neither sun, wind, or wet may enter, which would render the whole ineffectual.

Cleft-Grafting is performed by cleaving the stock. It is generally intended for large stocks, an inch or two diameter, cutting off the head at the height you design the stem, sloping one side about an inch in length, and with a strong knife, placed cross-ways at the top of the stock and sloped part, cleave it for the admission of the graft, wedging the cleft open until the graft is inserted; which is previously to be shortened to five or six inches, and the lower end sloped on two sides near two inches long; one edge made thicker than the other, and thereof the rind preserved entire, then

then introduce the graft into the back of the stock, with the thickest edge outwards, joining exactly rind to rind. Carefully remove the wedge, that the cleft may close upon the graft, and tye the parts firmly together, and clay them, as before directed.

If two grafts are to be inserted, cut off the head of the stock horizontally, cleave it right across, and insert a graft on each side, tying and claying them as before directed.

Crown-Grafting is performed occasionally upon stocks which are too large to cleave; the head being sawed or cut off horizontally, several grafts are inserted around the crown or top, betwixt the rind and wood; the grafts first sloped off on one side, forming a kind of shoulder at top, slit the rind, and separate it from the wood with a wedge, and introduce the grafts between, tie and clay them, as before.

Grafts and stocks in general effect their junction in six or eight weeks; and towards the end of May will begin to shoot, when the clay may be removed, but the bandage must remain three or four weeks longer; if the clay in any instance should crack or give way, immediately remove it, and apply more.

Budding or Inoculation.

This is effected by introducing small buds into the side of the stem or branch, between the bark and the wood, on young

young stocks about half an inch thick; in the next spring to be headed down to the budded part, when the bud will begin to push, and make a shoot, perhaps half a yard, or a yard in length.

The best season for budding is from the middle of July to the middle of August, as the buds should remain dormant till the following spring. If budded in June, they are apt to push out weak shoots the same year that will probably be killed in the ensuing winter. The buds for insertion are to be procured from young shoots of the same year's growth: procure therefore a quantity of cuttings, and take off their leaves to about a quarter of an inch of the buds; being furnished with a proper budding knife,

and

and strong new bass soaked in water, prepare the stock for the bud, at the height intended: on a smooth side of the stock cut the rind transversely, quite through to the wood, and from the middle of this make another cut downwards, an inch long, which, with the flat handle of the knife, must be opened on each side, separating the bark from the wood, then cut off one of the buds, enter the knife in the shoot, half an inch below the bud, cut a little into the wood, and run it slanting half an inch or more above, bringing it off with a small portion of the wood adhering, which must be directly detached from the bud, either with the point of the knife or the thumb and finger, observing instantly whether the eye or gem of the bud remains. If a small

small hole appears, it is bad, and another bud must be used; place it with the back part between the lips, till you have expeditiously opened the bark on each side of the perpendicular cut in the stock, clear to the wood; introduce it at top, slipping it down between the bark and wood, to the bottom, making the upper end also join with the horizontal cut at top; let the parts be immediately tyed with a ligature of fresh tough bass mat, bringing it closely round from bottom to top, except just over the eye of the bud, and the business is finished, no claying being wanted, as in grafting. In three or four weeks the buds will unite with the stock, and the parts begin to swell, then loosen the bandages. Just before they begin to shoot, which will not be till spring,

let

let the head of each stock be cut off aslanting, a little above the bud, which will soon push forth. In the Autumn the trees may be transplanted into the garden, or as directed in the respective articles.

Observe, that all shoots arising from the stock or stem, except the grafts or buds themselves, must always be displaced as soon as possible. Further particulars will be found fully explained under the proper articles.

Situation, Exposure, and Soil.

CONCERNING the situation eligible for *Fruit Trees*, it may be observed, that thriving trees and good fruit are produced both in high and low grounds, where the soil is proper; too low a situation,

ation however, is subject to inundations, or too copious moisture. Few fruit trees are ever prosperous for any continuance in very low wet places; however, in most situations, whether higher, lower, or moderately sloping, if good soil and not wet, most sorts of fruit trees may be successfully cultivated.

Sometimes a moderately low situation, not wet, may have the advantage, by being more out of the power of cutting blasts and tempestuous winds. A gentle slope towards the south, or south-east or west, is a desirable position. Where the situation for an orchard can be so contrived as to have the shelter of an adjacent forest tree plantation, at a little northerly distance, it will be an additional advantage.

As to Exposure, this may be varied in different trees. In wall-trees it may be more effectually practised by planting the more tender and choice forts against walls of a foutherly expofure. Walls of an East or West afpect are proper for fucceffional crops of the fame forts, as well for the more hardy and common wall-fruit, northerly walls may continue hardy fummer fruits, late in the feafon; fuch as cherries, plums, currants, &c.

Though an afpect full to the fun is always to be preferred, we may fee fruit-trees in a profperous ftate in almoft all expofures, even fometimes full to the North, the moft unfavourable of all.

With refpect to Soil, it may be alfo obferved, that moft fort of fruit-trees will profper in any common good foil,

being

being one good spade deep of fertile mold; but if two or three spades deep, the greater the advantage. In a loamy soil, not too strong and clayey, most fruits are prosperous; but a moderately light sandy loam, of a pliable texture, free and easy to work at all seasons, makes a desirable soil. The strongest loams, however, may be meliorated with good dung, coal ashes, sandy earth, and other light opening substances. In gardens, not naturally of a loamy soil, we sometimes, from a pasture, common, or field, procure a sufficiency to prepare the borders intended for wall trees, either wholly, or part, working it with store of dung, and a portion of the natural soil of the borders, one or two spades deep. But any good earth of a blackish, hazelly, or brown colour, moderately light, fatty, and pliant, a spade or two deep will be found sufficiently eligible for fruit-trees. — Strong, stubborn, rank clayey soils are bad, but may be improved by light open mixtures; such as composts of sandy earths, coal ashes, plenty of rotten dung, &c. at all opportunities.

A too light or sandy soil must be fertilized by plenty of good dung, and occasional applications of strong sh earthy composts, and cold wettish soils must also be occasionally mended with light warming ingredients, such as just mentioned. But in low wet soils the situations for fruit trees should either be raised proportionally with composts of dung and earths worked up with the natural soil, sufficient to have the roots distant from

from under-ground water, or there should be canals or drains contrived to carry off the redundant moisture.

When the good soil of the garden is naturally too shallow, that is, less than a spade deep, some of the bad soil below shou'd be excavated a foot or more deep, and a composition of good earth and dung brought in to fill up the place. But in the open ground, where standard fruit trees are intended, and the soil requires amendment, and it may be too expensive and troublesome to make a general improvement, the addition of compost, from five to ten feet diameter, and one deep, will be sufficient upon the spot where the tree is to stand.

The borders for wall and espalier trees may be from about three or four to eight feet

feet wide, according to the size of the garden; but it is of much importance to have wide borders, that the roots may have sufficient scope to spread.

Method of Planting Fruit Trees.

ONE general method serves for all the different sorts of fruit trees.

In taking them up for planting, the greatest precaution is necessary for raising them with as large a spread of roots as possible; the root may require trimming, so far as to retrench any maimed or decayed parts, but retaining carefully all the main horizontal ones, mostly at full length, except just tipping off the ends sloping on the under side, and reducing any very long stragglers. With regard to the head; if a tree of only one

one year old, having the firſt ſhoot from the graft or budding entire, let the whole remain at the time of planting until March, then headed down to a few eyes, as directed in their different articles; if an older tree, which has already been trained in the nurſery, and has formed a head of branches, prune out only any very irregular ſhoot or branch, not confiſtent with the general form of the head, and reduce long ſtragglers, but retain all the regular branches entire, at leaſt for the preſent; and thoſe that are uſually ſhortened in the common courſe of pruning, ſuch as peaches, nectarines, &c. may have it performed after they are planted, as ordered under their proper heads; but apples, pears, plums, cherries, and ſuch other trees as are not generally ſhortened

shortened, should have all their regular branches retained entire. In large trees, with very full heads, it may be proper to reduce the long and crowding branches, not only to lessen the head in some proportion to the root, but that the power of the wind may be checked till the tree is firmly rooted afresh.

The trees being ready, and the spaces marked out for planting, proceed to dig a round aperture for each tree, capacious enough to admit its full spread of roots every way, about one spade deep, so as, when planted, the uppermost ones may be only from three to six inches below the surface; then placing the tree in the middle with its roots spread around, trim in the earth, the finest mold first, the rest as it comes to hand, breaking all

large

large clods, shaking the tree upward by the stem, to make the earth settle close between the roots and fibres; and when the earth is all in, tread it gently first round the outside to settle the earth to the extreme roots, then gradually towards the stem, to fix the plant in its proper position, finishing with a small hollow at top for occasional waterings. As soon as planted let wall and espalier trees be nailed and fastened to the walls, &c. and the standards, if tall stems and largish heads, supported with stout stakes to keep them steady till firmly rooted in their new quarters. It may likewise be proper in winter-planting, or late in spring, to mulch the surface of the ground about the tenderer or choicer kinds, to the full width of the hole the tree stands in, and about two or three inches thick, which will be a defence from winter froſt, or drying winds, till they have taken good root. Likewiſe in dry light ſoils may give a pot of water to ſettle the earth, and prepare the tree for rooting afreſh, particularly in late ſpring planting, repeating it occaſionally, but by no means water ſo freely as to render the earth miry, but moderately moiſt, for, if continued too wet, it will chill and rot the tender fibres.

Protecting the Bloſſoms, and thinning young Wall Fruit.

PROTECTION of the Bloſſoms of peach, nectarine, and apricot wall trees in unfavourable ſprings, is effected different ways, but the moſt eligible is either by mats, or cuttings of evergreens. The mats are to be occaſionally nailed up before the trees, when there is an appearance

pearance of sharp frosty nights; if the frost continues long, and no sun, let them remain up in the day also, but must be removed at every favourable appearance of moderate weather. The cuttings of evergreens, as being always furnished with leaves, are often used; stick these between the branches, so as to cover and afford shelter to the blossoms and infant fruit, and permit them to remain constantly night and day, till the fruit is fairly set, and somewhat advanced in growth; which often proves more effectual than matting, and is attended with considerably less trouble.

Sometimes, in default of a sufficiency of mats or evergreens, we use the cuttings of hornbeam, or the branches of dried fern, either of which, retaining the old leaves, and being stuck between the branches, will afford shelter.

Thinning

Thinning young wall fruit is occasionally necessary in favourable springs, when the tree sets more fruit than it can afford room for, or than it is able to bring to perfection, which is common to apricots, peaches, and nectarines. This work should begin when the fruit is not larger than the end of the finger, be pursued with great care and regularity, selecting the largest, most promising, and best placed fruit, to stand for the crop, and thin out the worst; leaving the proper fruit in proportion to the strength of the shoots; that is, one or two on the weaker shoots, three on the middling, and not more than four or five on the stronger shoots, but no where too close together.

FINIS.

www.ingramcontent.com/pod-product-compliance
Lightning Source LLC
Chambersburg PA
CBHW020244240426

43672CB00006B/630